$$\left[C^{-1}\right]^T \int_0^l \left[P\right] p(x) \, dx = \left\{F^N\right\}$$

$$\begin{Bmatrix} F_i \\ M_i \\ F_j \\ M_j \end{Bmatrix} = \begin{bmatrix} 1 & 0 & -\frac{3}{\lambda^2} & \frac{2}{\lambda^3} \\ 0 & 1 & -\frac{2}{\lambda} & \frac{1}{\lambda^2} \\ 0 & 0 & \frac{3}{\lambda^2} & -\frac{2}{\lambda^3} \\ 0 & 0 & -\frac{1}{\lambda} & \frac{1}{\lambda^2} \end{bmatrix} \int_0^l \begin{Bmatrix} 1 \\ x \\ x^2 \\ x^3 \end{Bmatrix} p(x) \, dx$$

ANALYSIS OF PLATES

ANALYSIS OF PLATES

David McFarland
Bert L. Smith
Walter D. Bernhart

School of Engineering
Wichita State University

SPARTAN BOOKS
NEW YORK · WASHINGTON

International Library of Congress Catalog Card Number 75-133107
International Standard Book Number 0-87671-560-9

Printed in the United States of America.

Sole distributor in Great Britain, the British
Commonwealth, and the Continent of Europe:

The Macmillan Press Ltd.
4 Little Essex Street
London WC2R 3LF

CONTENTS

INTRODUCTION

In the analysis of structures, we frequently encounter structural components for which one dimension, referred to as the thickness h, is much smaller than the other dimensions, as shown in Fig. I.1.

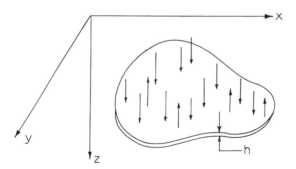

Fig. I.1. Plate

This type of structural component is called a plate. Several examples of plates which act as structural components are illustrated in Fig. I.2. Although plates are subjected to several different types of loads, in the material presented herein the loads applied to the plate are limited to those that act perpendicular to the plane of the plate, as shown in Fig. I.1.

The first significant analyses of plates began in the early 1800s, with much of the work attributed to Cauchy, Poisson, Navier, Lagrange, and Kirchhoff. The research done by these early "engineers" was extremely significant, and many of the techniques which they developed are still used in engineering analysis today. However, in the last 30

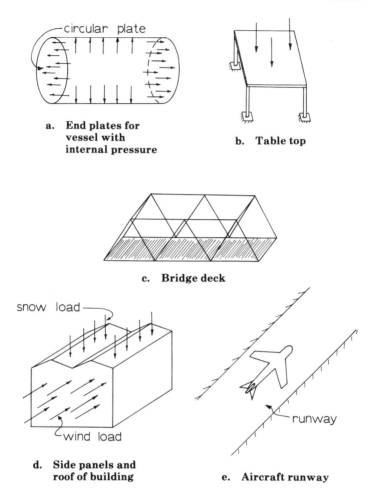

a. End plates for
   vessel with   b. Table top
   internal pressure

c. Bridge deck

d. Side panels and
   roof of building   e. Aircraft runway

Fig. I.2. Structural plates

years, with the advent of the digital computer, other methods such as finite differences and finite elements have become practical. The most recent and one of the most potentially powerful techniques for the analysis of plates and other structures is the technique of finite elements. We are just now beginning the second decade in which sufficient information is available to apply the method of finite elements to the analysis of plates.

Several books are available which deal in general with the analysis of plates. There are also many published papers, each of which presents an in depth study of a specific area of plate analysis. However, the authors have prepared this manuscript because they feel there is no individual work that serves by itself as a text for a course in thin plates, which covers derivations of basic relationships, classical solutions, and computer oriented solutions. This text is intended for use in a senior or graduate level course in the analysis of thin plates for stresses, strains, and lateral deflections.

The material presented herein, which is primarily concerned with small lateral deflections of thin plates subjected to lateral loads, is based upon the following four assumptions.

1. The middle plane of the plate remains unstrained during bending.

2. The effect of transverse shear strain is negligible.

3. Normal stresses in a direction perpendicular to the plane of the plate are neglected.

4. The material is homogeneous, isotropic, continuous, and linearly elastic.

Any exceptions to these assumptions are clearly pointed out. One notable exception is in Chapters 6 and 9 in which anisotropic plates are considered. Another is in Chapter 9 where the topic of large deflections of plates is presented.

The text is conveniently divided into three major parts:

Part I. Basic Relationships (Chapters 1 and 2)

Part II. Classical Solutions (Chapters 3 through 7)

Part III. Computer Oriented Solutions (Chapters 8 and 9)

In Chapter 1, the governing equation and boundary conditions defining the small lateral deflections of the middle surface of a thin rectangular plate are developed in terms of the Cartesian coordinates x and y. To completely describe the behavior of the plate, expressions relating the lateral deflection to displacements, strains, and stresses are also developed. Thus, the solution of the governing equation for the lateral deflection gives all the information necessary to determine displacements, strains, and stresses. Consequently, the emphasis of this text is on the determination of the lateral deflection for plates of various shapes and boundary conditions. In Chapter 2, similar expressions are derived for circular plates in terms of the polar coordinates r and θ by use of the transformation which relates polar coordinates to Cartesian coordinates.

Lateral deflections of rectangular plates are determined in Chapter 3 by the classical methods of Levy and Navier. Also in Chapter 3, the principle of superposition combined with the Levy method is used to determine deflections of rectangular plates with various boundary conditions. A classical solution for determining deflections of circular plates is presented in Chapter 4. Deflections of continuous plates supported by intermediate beams and columns are discussed in Chapter 5, in addition to plates on elastic foundations. In Chapter 6, an exception is made to Assumption 4, and deflections of orthotropic plates are considered. Energy methods are presented in Chapter 7 where particular attention is given to the method of Ritz and the application of Lagrange multipliers.

Chapters 8 and 9 deal with the approximate techniques of finite differences and finite elements which are used to determine deflections of plates of various shapes and boundary conditions. These techniques, particularly finite elements, are highly computer oriented. In most cases, the quantity of numerical manipulations required for these techniques would be prohibitive without the aid of a digital computer. However, since most engineers have digital computation facilities at their disposal, these techniques are very valuable and highly practical. Included in Chapter 9 are the special topics of laminated plates, orthotropic plates, and large deflections of plates.

This text is organized so that Chapter 1 must be studied first. After Chapter 1 has been studied, the remaining chapters may be taken in any sequence with the exception of Chapter 4 (Chapter 2 must be studied before Chapter 4 is considered). However, the authors have arranged the chapters in a sequence compatible with an orderly study of the analysis of plates. Parts II and III contain many example problems to demonstrate the application of the theory presented. The authors wish to emphasize the importance of the study of these examples for a more thorough understanding of the subject. The text is suitable for either a one semester or a two semester course.

Basic Relationships

Chapter 1
BASIC RELATIONSHIPS FOR
RECTANGULAR PLATES

1.1 ASSUMPTIONS

The equation defining small lateral deflections of the middle surface of a thin plate subjected to lateral loads may be formulated in different ways. The most general method involves the elimination of unimportant terms from the equations of three-dimensional elasticity as the thickness is made small compared with other dimensions. A second method, which requires less mathematical rigor but more physical interpretation, utilizes the basic assumptions made in the theory of beams to generate directly the equations for thin plates. Both methods lead to the same results. The second approach is taken here, since it is considered more appealing to the engineer. The development of the equations from this approach is based on the following four assumptions.

1. The middle surface of the plate remains unstrained during bending; thus, it is a neutral surface.

2. Normals to the middle surface before deformation remain normal to the same surface after deformation. As in simple beam theory, this assumption does not imply that the transverse shear strain is necessarily zero. The assumption implies that the transverse shear is so small that any distortion of transverse sections caused by the existence of transverse shear strain makes a negligible contribution to displacements. It should be noted that even for thin plates this assumption leads to unreliable results in certain edge regions such as those near

allows one to neglect effects of transverse shears

only lateral loads are small deflections

3

corners and near holes with diameters of the order of magnitude of the plate thickness.

3. Normal stresses in the direction transverse to the plate are small compared with other stresses; thus, they are neglected. This assumption becomes unreliable even for thin plates in the neighborhood of highly concentrated transverse loads.

4. The material is homogeneous, isotropic, continuous, and linearly elastic. This assumption, which is an idealization on the material properties, permits the use of the stress-strain relationships in terms of two elastic constants.

Large deflections[1] generally cause excessive middle surface strain in violation of Assumption 1. An exception is the plate whose middle surface can be bent into a developable surface, which is a surface that can be rolled out on a plane without changes in the distances between any two points on the surface. Examples are surfaces of cones and cylinders. If the middle surface is a developable surface, it remains unstrained even for large deflections. Large deflections also give rise to nonlinear equations which are more difficult to solve. Plates of considerable thickness[1-3] require solution by a more general theory because Assumptions 2 and 3 become unreliable.

A precise definition for "thin" and "small" must be obtained for each problem individually by comparing the results obtained from the thin plate small deflection theory to those results obtained from a more exact theory. A precise definition is not always practical. However, evidence shows that the assumptions stated here provide the basis for a reliable theory for plates having a maximum deflection of the order of magnitude of their thickness, and a thickness up to approximately five percent of the other dimensions such as diameter, length, and width. Remember, this is a generalization to introduce the reader to the concept of "thin" and "small," and by no means should be interpreted as a rigid rule.

These four assumptions permit the derivation of the governing differential equation for the deflection of the middle surface of the plate in terms of transverse applied loads to be accomplished in a concise, straightforward manner. At key places in the various derivations presented in the following sections, reference is made to the appropriate assumption upon which that derivation or that portion of the derivation is based.

1.2   COORDINATES AND DIMENSIONS

The x and y axes lie in the middle plane of the plate before deformation and are parallel to the edges, and the z axis is perpendicular to the plate as shown in Fig. 1.1.

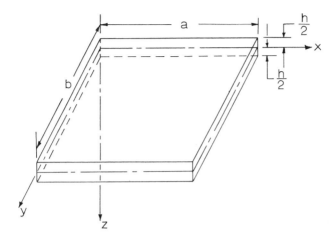

Fig. 1.1. Rectangular plate coordinates and dimensions

The perpendicular distance of a point in the plate measured from the middle surface is given by the coordinate ζ and should not be confused with the coordinate z or the lateral deflection w of the middle surface of the plate. The distinction between these coordinates and the deflection of the middle surface is shown in Fig. 1.2. The thickness of the plate is h; thus, the extremes of ζ are $\pm h/2$.

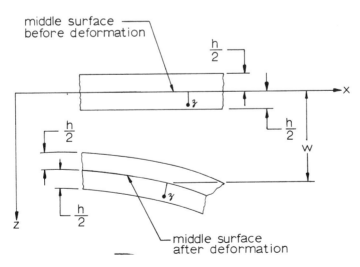

Fig. 1.2. Projection of transverse section of plate on x-z plane showing distinction between z, ζ, and w

1.3 DISPLACEMENTS

The displacements in the x, y, and z directions respectively are u, v, and w. Under transverse loads, the middle surface deforms into a curved surface defined by the equation

$$w = f(x, y),$$

where

    w = lateral deflection of the middle surface in the z direction from the x-y plane as shown in Figs. 1.2 and 1.3,

    x, y = coordinates of a point on the middle surface.

The displacement in the x and y directions of a point at a distance $\mathfrak{z}$ from the middle surface are u and v respectively. These displacements are illustrated in Fig. 1.3, which shows the middle surface before and after deformation. Notice that the normals to the middle surface before deformation remain normal to the middle surface after deformation according to Assumption 2. Since the existence of transverse shear strain is possible, even though it is considered to be small as compared to other strains, a normal to the middle surface before deformation does not really remain normal to the middle surface after deformation. It is distorted from its normal position by a small angle which is defined as the transverse shear strain. An example of this distortion is shown

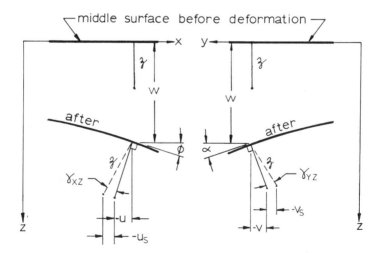

a. Displacement in x direction, u.
Note, u_s as caused by γ_{xz} is neglected.

b. Displacement in y direction, v.
Note, v_s as caused by γ_{yz} is neglected.

Fig. 1.3. Displacements of a point in a plate a distance $\mathfrak{z}$ from the middle surface

by the dashed lines in Fig. 1.3. It is assumed that any contributions to the displacements u and v from these distortions, illustrated by u_s and v_s in Fig. 1.3, are negligible. The displacements u and v are assumed to exist in their entirety because of kinematic rotations of the normals. However, it is important to remember that the disregard of the effect of transverse shear on the displacements does not preclude its existence, as will be shown in Section 1.8.

Expressions for the displacements u and v in terms of the deflection w(x,y) of the middle surface are determined from trigonometric consideration as follows.

$$u = -\beta \sin \phi, \qquad v = -\beta \sin \alpha$$

If the deflections are small,

$$\sin \phi = \frac{\partial w}{\partial x} \quad \text{and} \quad \sin \alpha = \frac{\partial w}{\partial y}. \tag{1.1}$$

Thus, the displacements become

$$u = -\beta \frac{\partial w}{\partial x}$$

$$v = -\beta \frac{\partial w}{\partial y}. \tag{1.2}$$

Notice that the displacements u and v evaluated at the middle surface vanish according to Assumption 1.

1.4 STRAINS

The normal strains in the x and y directions, ϵ_x and ϵ_y, and the associated shear strain, γ_{xy}, which are in the plane that is located at an arbitrary perpendicular distance β from the middle surface, can be expressed in terms of the deflection of the middle surface w(x,y). The development of these expressions is based on the criterion of a neutral middle surface for which normals remain normal as stated by Assumptions 1 and 2. Fig. 1.4 shows a surface of a plate in the x-z plane after bending.

From geometry

$$ds_x = \rho_x d\phi_x \quad \text{and} \quad ds'_x = (\rho_x - \beta) d\phi_x.$$

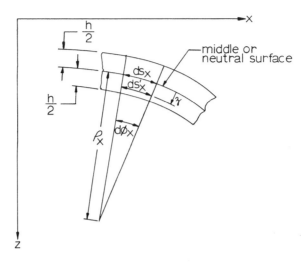

Fig. 1.4.   Surface of a plate in the x-z plane after bending

The expression for the normal strain in the x direction at a distance $\mathfrak{z}$ from the middle surface is, by definition,

$$\epsilon_x = -\frac{ds_x - ds_x'}{ds_x} = -\mathfrak{z}\frac{d\phi_x}{ds_x} = -\frac{\mathfrak{z}}{\rho_x}. \tag{1.3}$$

Before the normal strain ϵ_x can be expressed in terms of the deflection of the middle surface, an expression for the radius of curvature ρ_x as a function of the deflection of the middle surface must first be developed. Fig. 1.5 shows a projection of the deflected middle surface on the x-z

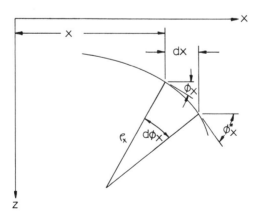

Fig. 1.5.   Projection of the middle surface on the x-z plane and its radius of curvature, ρ_x

plane and the radius of curvature ρ_x associated with this projected curve.
If the deflections $w(x,y)$ are small

$$ds_x = dx$$

$$\phi_x = \frac{\partial w}{\partial x}$$

$$\phi_x^* = \phi_x + \frac{\partial \phi_x}{\partial x} dx = \frac{\partial w}{\partial x} + \frac{\partial^2 w}{\partial x^2} dx$$

$$d\phi_x = \phi_x^* - \phi_x = \frac{\partial^2 w}{\partial x^2} dx$$

and

$$\frac{d\phi_x}{dx} = \frac{\partial^2 w}{\partial x^2}.$$

The expression for the slope of the middle surface at $x + dx$ is obtained by expanding it into a Taylor's series about x. The higher-order terms are considered negligible since dx is a quantity of infinitesimal value.

Again, from geometry

$$ds_x = \rho_x d\phi_x$$

or
$$\frac{1}{\rho_x} = \frac{d\phi_x}{ds_x} = \frac{d\phi_x}{dx} = \frac{\partial^2 w}{\partial x^2}. \tag{1.4a}$$

Similarly

$$\frac{1}{\rho_y} = \frac{\partial^2 w}{\partial y^2}. \tag{1.4b}$$

Substitution of Eq. (1.4a) into (1.3) permits the normal strain in the x direction to be expressed in terms of the deflection of the middle surface.

$$\epsilon_x = -\mathfrak{z} \frac{\partial^2 w}{\partial x^2} \tag{1.5a}$$

Similarly

$$\epsilon_y = -\mathfrak{z}\,\frac{\partial^2 w}{\partial y^2}. \qquad (1.5b)$$

Because of Assumption 3, which asserts that the normal stress σ_z may be neglected, it will be shown in the next section that the normal strain ϵ_z can be expressed in terms of the other two components of normal strain, ϵ_x and ϵ_y.

To obtain the corresponding expression for the shear strain, a plane differential rectangular element of the plate in the x-y plane a distance $\mathfrak{z}$ from the middle surface, as shown in Fig. 1.6, is considered. The un-

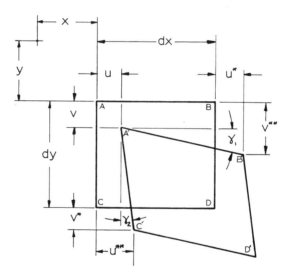

Fig. 1.6. Deformation of a differential element on a plane at a distance $\mathfrak{z}$ from the middle surface

strained position of the element is represented by position ABCD which is in an x-y plane. After strain, which is caused by deflection of the middle surface of the plate, the element is displaced to position A′B′C′D′. For small deflections of the middle surface, the strains are determined from the consideration that the strained position A′B′C′D′ remains an x-y plane. There are two basic types of geometric deformation. One type is a linear deformation or a change in the lengths of the sides of the element; the other type is an angular deformation or a change in the value of a given angle. Appreciable changes are considered to take place in the x-y plane only.

Expressions for the components of linear displacement at B(x + dx,y) and $C(x, y + dy)$ are obtained by expanding each into a Taylor's series about $A(x,y)$. The higher-order terms are considered negligible since dx and dy are quantities of infinitesimal value.

$$u^* = u + \frac{\partial u}{\partial x}\, dx$$

$$v^* = v + \frac{\partial v}{\partial y}\, dy$$

$$u^{**} = u + \frac{\partial u}{\partial y}\, dy$$

$$v^{**} = v + \frac{\partial v}{\partial x}\, dx$$

The shear strain is defined as the angular deformation of a given angle. The angle CAB of Fig. 1.6 deforms to the angle $C'A'B'$, the deformation being the angle $\gamma_1 + \gamma_2$; thus, the shear strain is

$$\gamma_{xy} = \gamma_1 + \gamma_2.$$

For small deformations

$$\gamma_1 = \frac{v^{**} - v}{dx} = \frac{\partial v}{\partial x}$$

and

$$\gamma_2 = \frac{u^{**} - u}{dy} = \frac{\partial u}{\partial y}.$$

Thus, the expression for shear strain in terms of displacements is

$$\gamma_{xy} = \frac{\partial u}{\partial y} + \frac{\partial v}{\partial x}. \tag{1.6}$$

The expression for the shear strain in terms of the deflection of the middle surface, which can be obtained by substituting Eqs. (1.1) and (1.2) into Eq. (1.6), is

$$\gamma_{xy} = -2\, \mathfrak{z}\, \frac{\partial^2 w}{\partial x\, \partial y}. \tag{1.7a}$$

The expressions for the normal strains given by Eqs. (1.5) can also be obtained from Fig. 1.6. By definition of normal strain

$$\epsilon_x = \frac{A'B' - AB}{AB} = \frac{u^* - u}{dx} = \frac{\partial u}{\partial x} = -\vartheta \frac{\partial^2 w}{\partial x^2} \qquad (1.7b)$$

and

$$\epsilon_y = \frac{A'C' - AC}{AC} = \frac{v^* - v}{dy} = \frac{\partial v}{\partial y} = -\vartheta \frac{\partial^2 w}{\partial y^2}. \qquad (1.7c)$$

It is shown in the next section that expressions for the transverse shear strains γ_{xz} and γ_{yz} are not required. Their presence is required for the more general plate theory[1,4] that involves the effect of transverse shear deformation which is not considered here.

1.5   STRESSES

The stress-strain relationships used here are those which can be expressed in terms of two independent engineering constants: the modulus of elasticity, E; and Poisson's ratio, ν. The derivation of these relationships is based on an idealization of the material properties.[5] The material is considered to be homogeneous, isotropic, continuous, and linearly elastic according to Assumption 4. Familiarity with these relationships is assumed; thus, without derivation, they are

$$\epsilon_x = \frac{1}{E} \left(\sigma_x - \nu\sigma_y - \nu\sigma_z \right) \qquad (1.8a)$$

$$\epsilon_y = \frac{1}{E} \left(-\nu\sigma_x + \sigma_y - \nu\sigma_z \right) \qquad (1.8b)$$

$$\epsilon_z = \frac{1}{E} \left(-\nu\sigma_x - \nu\sigma_y + \sigma_z \right) \qquad (1.8c)$$

$$\gamma_{xy} = \frac{2(1 + \nu)}{E} \tau_{xy} \qquad (1.8d)$$

$$\gamma_{xz} = \frac{2(1 + \nu)}{E} \tau_{xz} \qquad (1.8e)$$

$$\gamma_{yz} = \frac{2(1 + \nu)}{E} \tau_{yz}. \qquad (1.8f)$$

According to Assumption 3, the normal stress σ_z in the direction transverse to the plate is small compared with other stresses. This assumption allows Eqs. (1.8) to be written in the following form.

$$\sigma_x = \frac{E}{1 - \nu^2} (\epsilon_x + \nu \epsilon_y) \tag{1.9a}$$

$$\sigma_y = \frac{E}{1 - \nu^2} (\nu \epsilon_x + \epsilon_y) \tag{1.9b}$$

$$\sigma_z = \text{negligible}$$

or

$$\sigma_x + \sigma_y = -\frac{E}{\nu} \epsilon_z \tag{1.9c}$$

$$\tau_{xy} = \frac{E}{2(1 + \nu)} \gamma_{xy} \tag{1.9d}$$

$$\tau_{xz} = \frac{E}{2(1 + \nu)} \gamma_{xz} \tag{1.9e}$$

$$\tau_{yz} = \frac{E}{2(1 + \nu)} \gamma_{yz} \tag{1.9f}$$

As mentioned previously, the transverse shear stresses of Eqs. (1.9e) and (1.9f) exist; however, their expressions are not required for the present development of the equation defining the lateral deflection of the middle surface. The stresses of Eqs. (1.9a), (1.9b), and (1.9d) can now be written in terms of the deflection of the middle surface by referring to Eqs. (1.5a), (1.5b), and (1.7).

$$\sigma_x = -\jmath \frac{E}{1 - \nu^2} \left(\frac{\partial^2 w}{\partial x^2} + \nu \frac{\partial^2 w}{\partial y^2} \right) \tag{1.10a}$$

$$\sigma_y = -\jmath \frac{E}{1 \oplus \nu^2} \left(\nu \frac{\partial^2 w}{\partial x^2} + \frac{\partial^2 w}{\partial y^2} \right) \tag{1.10b}$$

$$\tau_{xy} = -\jmath \frac{E}{1 + \nu} \frac{\partial^2 w}{\partial x \, \partial y} \tag{1.10c}$$

1.6   STRESS RESULTANTS

In the previous section, the plate stresses were discussed and expressions were derived for them in terms of the deflection of the middle surface. The positive state of stress can be shown by considering the

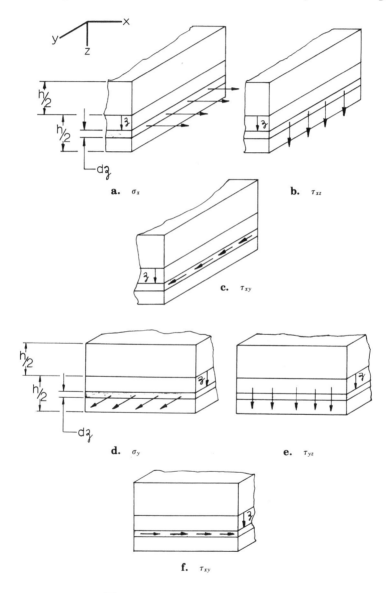

Fig. 1.7.   Positive state of stress

action of the stresses on a lamina of a rectangular element as shown in Fig. 1.7. To maintain clarity, the plate element is shown in a series of six figures. Each figure illustrates the action of a different stress. On the two transverse faces of the element shown, the positive stresses are in the directions of the positive coordinate axes. On the opposite two transverse faces, not shown, positive stresses are assumed to be in the directions of the negative coordinate axes.

The stress distribution on the transverse faces of the element can be reduced to moments and forces per unit of length in the x and y directions. These moments and forces are defined as stress resultants. The positive state of stress resultants is shown on the rectangular element in Fig. 1.8. On the two transverse faces of the element not shown, the stress resultants are considered positive if they act in the opposite directions as those shown on the transverse faces of the elements in Fig. 1.8.

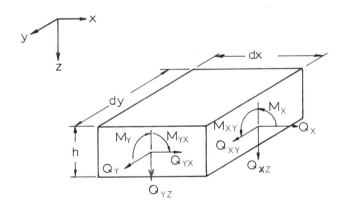

Fig. 1.8. Positive state of stress resultants

According to the definition stated for stress resultants, they are expressed as follows.

$$M_x = \int_{-h/2}^{h/2} \mathfrak{z}\, \sigma_x \, d\mathfrak{z}$$

$$M_y = \int_{-h/2}^{h/2} \mathfrak{z}\, \sigma_y \, d\mathfrak{z}$$

$$M_{xy} = M_{yx} = - \int_{-h/2}^{h/2} \mathfrak{z} \, \tau_{xy} \, d\mathfrak{z}$$

$$Q_{xy} = Q_{yx} = \int_{-h/2}^{h/2} \tau_{xy} \, d\mathfrak{z}$$

$$Q_{xz} = \int_{-h/2}^{h/2} \tau_{xz} \, d\mathfrak{z}$$

$$Q_{yz} = \int_{-h/2}^{h/2} \tau_{yz} \, d\mathfrak{z}$$

$$Q_x = \int_{-h/2}^{h/2} \sigma_x \, d\mathfrak{z}$$

$$Q_y = \int_{-h/2}^{h/2} \sigma_y \, d\mathfrak{z}$$

Integration of these expressions for the stress resultants after substitution of Eqs. (1.10) leads to the following expressions.

$$M_x = -D \left(\frac{\partial^2 w}{\partial x^2} + \nu \frac{\partial^2 w}{\partial y^2} \right) \qquad (1.11a)$$

$$M_y = -D \left(\nu \frac{\partial^2 w}{\partial x^2} + \frac{\partial^2 w}{\partial y^2} \right) \qquad (1.11b)$$

$$M_{xy} = M_{yx} = D(1 - \nu) \frac{\partial^2 w}{\partial x \, \partial y} \qquad (1.11c)$$

$$Q_x = Q_y = Q_{xy} = Q_{yx} = 0 \qquad (1.11d)$$

where

$$D = \frac{Eh^3}{12(1 - \nu^2)}.$$

Expressions for the transverse shear stress resultants Q_{xz} and Q_{yz} in terms of the deflection $w(x,y)$ will be developed from equilibrium considerations as presented in the next section.

1.7   EQUILIBRIUM OF A DIFFERENTIAL PLATE ELEMENT

The condition of equilibrium is met by considering a rectangular differential element of dimensions dx, dy, and h as shown in Fig. 1.9. For simplicity, only the middle surface of the plate is shown. The stress resultants, which are internal reactions per unit of length, exist because of the transverse distributed load of intensity p(x,y) on the upper surface of the plate.

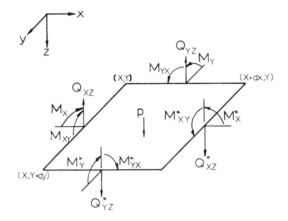

Fig. 1.9.   Differential plate element (middle surface) with stress resultants

Expressions for the stress resultants acting on the transverse faces at (x + dx,y) and at (x,y + dy) are obtained by expanding each into a Taylor's series about (x,y). The higher-order terms are considered negligible since dx and dy are quantities of infinitesimal value.

$$Q_{yz}^* = Q_{yz} + \frac{\partial Q_{yz}}{\partial y}\, dy$$

$$Q_{xz}^* = Q_{xz} + \frac{\partial Q_{xz}}{\partial x}\, dx$$

$$M_y^* = M_y + \frac{\partial M_y}{\partial y}\, dy$$

$$M_x^* = M_x + \frac{\partial M_x}{\partial x}\, dx$$

$$M_{yx}^* = M_{yx} + \frac{\partial M_{yx}}{\partial y}\, dy$$

$$M_{xy}^* = M_{xy} + \frac{\partial M_{xy}}{\partial x}\, dx$$

The condition of a vanishing resultant force in the z direction results in the equation

$$-Q_{xz}\, dy - Q_{yz}\, dx + Q_{yz}^*\, dx + Q_{xz}^*\, dy + p\,dxdy = 0$$

or
$$-Q_{xz}\, dy - Q_{yz}\, dx + \left(Q_{yz} + \frac{\partial Q_{yz}}{\partial y}\, dy \right) dx$$
$$+ \left(Q_{xz} + \frac{\partial Q_{xz}}{\partial x}\, dx \right) dy + p\,dxdy = 0$$

or
$$\frac{\partial Q_{yz}}{\partial y} + \frac{\partial Q_{xz}}{\partial x} + p = 0. \tag{1.12a}$$

If the resultant moment about an edge parallel to the x axis is set equal to zero while neglecting higher-order terms, the resulting equation is

$$\frac{\partial M_y}{\partial y}\, dxdy - \frac{\partial M_{xy}}{\partial x}\, dxdy - Q_{yz}\, dxdy = 0$$

or
$$\frac{\partial M_y}{\partial y} - \frac{\partial M_{xy}}{\partial x} - Q_{yz} = 0. \tag{1.12b}$$

Similarly, the equilibrium equation with respect to rotation about an edge parallel to the y axis is

$$\frac{\partial M_x}{\partial x} - \frac{\partial M_{yx}}{\partial y} - Q_{xz} = 0. \tag{1.12c}$$

If the expression for Q_{yz} from Eq. (1.12b) and the expression for Q_{xz} from Eq. (1.12c) is substituted into Eq. (1.12a), the resulting equation is

$$\frac{\partial^2 M_x}{\partial x^2} - 2\frac{\partial^2 M_{xy}}{\partial x\, \partial y} + \frac{\partial^2 M_y}{\partial y^2} = -p. \tag{1.13}$$

It was stated in the previous section that the development of expressions for the transverse stress resultants Q_{xz} and Q_{yz} would be done at this time. Substitution of Eqs. (1.11a), (1.11b), and (1.11c) into Eqs. (1.12b) and (1.12c) gives the expressions for Q_{xz} and Q_{yz} in terms of the deflection of the middle surface

$$Q_{xz} = -D \left(\frac{\partial^3 w}{\partial x^3} + \frac{\partial^3 w}{\partial x \, \partial y^2} \right) = -D \frac{\partial}{\partial x} (\nabla^2 w) \qquad (1.14a)$$

and

$$Q_{yz} = -D \left(\frac{\partial^3 w}{\partial y^3} + \frac{\partial^3 w}{\partial x^2 \, \partial y} \right) = -D \frac{\partial}{\partial y} (\nabla^2 w) \qquad (1.14b)$$

where
$$\nabla^2 = \frac{\partial^2}{\partial x^2} + \frac{\partial^2}{\partial y^2}.$$

1.8   GOVERNING EQUATION

The governing partial differential equation defining the lateral deflection of the middle surface of the plate in terms of the applied transverse load is obtained by direct substitution of Eqs. (1.11a), (1.11b), and (1.11c) into the equilibrium Eq. (1.13). The result of this is

$$\frac{\partial^4 w}{\partial x^4} + 2 \frac{\partial^4 w}{\partial x^2 \, \partial y^2} + \frac{\partial^4 w}{\partial y^4} = \frac{p}{D} \qquad (1.15a)$$

or
$$\nabla^4 w = \frac{p}{D} \qquad (1.15b)$$

or
$$\nabla^2 \nabla^2 w = \frac{p}{D} \qquad (1.15c)$$

where
$$\nabla^4 = \frac{\partial^4}{\partial x^4} + 2 \frac{\partial^4}{\partial x^2 \, \partial y^2} + \frac{\partial^4}{\partial y^4}.$$

The fourth-order partial differential Eq. (1.15) can be reduced to two second-order partial differential equations which are sometimes preferred, depending upon the method of solution to be used. This reduction, first discussed by Marcus,[10] is accomplished as follows. The addition of Eqs. (1.11a) and (1.11b) gives

$$M_x + M_y = -D(1 + \nu)\left(\frac{\partial^2 w}{\partial x^2} + \frac{\partial^2 w}{\partial y^2}\right)$$

or

$$\frac{\partial^2 w}{\partial x^2} + \frac{\partial^2 w}{\partial y^2} = -\frac{M_x + M_y}{D(1 + \nu)}$$

or

$$\nabla^2 w = \frac{M}{D} \qquad\qquad (1.16a)$$

where

$$M = -\frac{M_x + M_y}{1 + \nu}.$$

Substitution of Eq. (1.16a) into (1.15c) gives

$$\nabla^2\left(\frac{M}{D}\right) = \frac{p}{D}$$

or

$$\nabla^2 M = p. \qquad\qquad (1.16b)$$

Thus, the fourth-order Eq. (1.15) has been reduced to the two second-order Eqs. (1.16a) and (1.16b). If boundary conditions and the transverse load p are known, Eq. (1.16b) can be solved for $M(x, y)$. Then, Eq. (1.16a) can be solved for $w(x, y)$.

1.9   BOUNDARY CONDITIONS

A complete solution of the governing Eq. (1.15) depends upon the knowledge of the conditions of the plate at the boundaries in terms of the lateral deflection of the middle surface $w(x, y)$. Thus, expressions for these conditions must be developed. Three types of boundaries are considered at this time: simply supported, clamped, and free.

Simple Supported Edge Conditions. A plate boundary that is prevented from deflecting but free to rotate about a line along the boundary edge, such as a hinge, is defined as a simply supported edge. The conditions on a simply supported edge parallel to the y axis at $x = a$, Fig. 1.1, are

$$w\Big|_{x=a} = 0$$

$$M_x\Big|_{x=a} = -D\left(\frac{\partial^2 w}{\partial x^2} + \nu\frac{\partial^2 w}{\partial y^2}\right)_{x=a} = 0.$$

Since the change of w with respect to the y coordinate vanishes along this edge, these conditions become

$$w \Big|_{x = a} = 0 \qquad (1.17a)$$

$$\frac{\partial^2 w}{\partial x^2} \Big|_{x = a} = 0. \qquad (1.17b)$$

On a simply supported edge parallel to the x axis at y = b, Fig. 1.1, the change of w with respect to the x coordinate vanishes; thus, the conditions along this boundary are

$$w \Big|_{y = b} = 0 \qquad (1.18a)$$

$$M_y \Big|_{y = b} = -D \left(\nu \frac{\partial^2 w}{\partial x^2} + \frac{\partial^2 w}{\partial y^2} \right)_{y = b}$$

$$= -D \frac{\partial^2 w}{\partial y^2} \Big|_{y = b} = 0. \qquad (1.18b)$$

Clamped Edge Conditions. If a plate boundary is clamped, the deflection and the slope of the middle surface must vanish at the boundary. On a clamped edge parallel to the y axis at x = a, Fig. 1.1, the boundary conditions are

$$w \Big|_{x = a} = 0 \qquad (1.19a)$$

$$\frac{\partial w}{\partial x} \Big|_{x = a} = 0. \qquad (1.19b)$$

The boundary conditions on a clamped edge parallel to the x axis at y = b are

$$w \Big|_{y = b} = 0 \qquad (1.20a)$$

$$\frac{\partial w}{\partial y} \Big|_{y = b} = 0. \qquad (1.20b)$$

Free Edge Conditions. In the most general case, a twisting moment, a bending moment, and a transverse shear force act on an edge of a plate, as shown in Fig. 1.10.

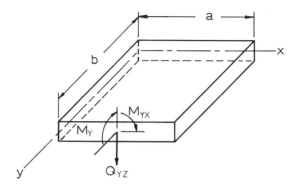

Fig. 1.10. Stress resultants at y = b

An edge on which all three of these stress resultants vanish is defined as a free edge. Two boundary conditions at each edge are sufficient for a complete solution of the governing Eq. (1.15); thus, the three conditions presented here are too many. The inconsistency is created because of Assumption 2, which permits the disregard of the effect of transverse shear strain for the determination of the expressions for displacements. Without this assumption, a sixth-order equation would be obtained rather than a fourth-order equation, and three boundary conditions would then be required, as shown by Reissner.[4] Based on the theorem that the first variation of the potential energy must vanish for an equilibrium configuration, Kirchhoff[6] first showed that two boundary conditions on a free edge are sufficient for thin plate theory. The sufficiency of two boundary conditions on a free edge was also shown by Rayleigh[8] using the same technique, which is illustrated in Chapter 7.

One of the boundary conditions derived by Kirchhoff for a free edge corresponds to equating the bending moment stress resultant to zero. The other boundary condition corresponds to setting equal to zero an expression involving both the twisting moment stress resultant and the transverse shear stress resultant. The physical significance of the latter of these two boundary conditions was explained by Thomson and Tait.[7] An elaboration of their explanation is presented here.

The twisting moment stress resultant M_{yx}, Fig. 1.11a, is replaced by equivalent couples of equal and opposite vertical forces as shown in Fig. 1.11b. The components M_{yx} and $M_{yx} + (\partial M_{yx}/\partial x)\,dx$ oppose each other and, thus, combine to form one vertical component of magnitude $(\partial M_{yx}/\partial x)\,dx$ as shown in Fig. 1.11c. The component $(\partial M_{yx}/\partial x)\,dx$ combines with the shear stress resultant Q_{yz} to give the equivalent transverse shear stress resultant V_{yz}.

$$V_{yz}\,dx\,\bigg|_{y\,=\,b} = Q_{yz}\,dx\,\bigg|_{y\,=\,b} - \frac{\partial M_{yx}}{\partial x}\,dx\,\bigg|_{y\,=\,b}$$

or

$$V_{yz}\,\bigg|_{y\,=\,b} = \left(Q_{yz} - \frac{\partial M_{yx}}{\partial x}\right)_{y\,=\,b}$$

or

$$V_{yz}\,\bigg|_{y\,=\,b} = -D\left(\frac{\partial^3 w}{\partial y^3} + (2 - \nu)\frac{\partial^3 w}{\partial x^2\,\partial y}\right)_{y\,=\,b} \qquad (1.21a)$$

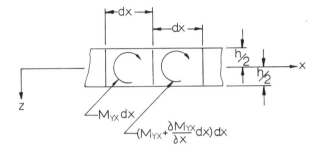

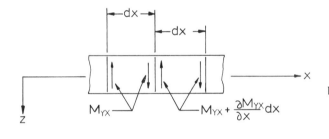

a. Twisting moments

b. Twisting moments represented by equivalent vertical forces

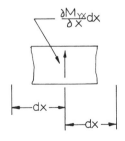

c. Twisting moment represented by equivalent vertical force

Fig. 1.11. Edge of plate, at y = b, under action of twisting moments

and similarly

$$V_{xz}\Big|_{x\,=\,a} = -D\left(\frac{\partial^3 w}{\partial x^3} + (2 - \nu)\frac{\partial^3 w}{\partial x\,\partial y^2}\right)_{x\,=\,a}. \quad (1.21b)$$

These reduced stress resultants are shown on the differential element in Fig. 1.12.

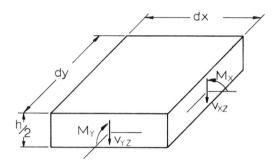

Fig. 1.12. Positive state of reduced stress resultants for free edges

The boundary conditions on a free edge parallel to the x axis at $y = b$, Fig. 1.1, are

$$M_y\Big|_{y\,=\,b} = -D\left(\nu\frac{\partial^2 w}{\partial x^2} + \frac{\partial^2 w}{\partial y^2}\right)_{y\,=\,b} = 0 \quad (1.22a)$$

and $\quad V_{yz}\Big|_{y\,=\,b} = -D\left(\frac{\partial^3 w}{\partial y^3} + (2 - \nu)\frac{\partial^3 w}{\partial x^2\,\partial y}\right)_{y\,=\,b} = 0. \quad (1.22b)$

The boundary conditions on a free edge parallel to the y axis at $x = a$ are

$$M_x\Big|_{x\,=\,a} = -D\left(\frac{\partial^2 w}{\partial x^2} + \nu\frac{\partial^2 w}{\partial y^2}\right)_{x\,=\,a} = 0 \quad (1.23a)$$

$$V_{xz}\Big|_{x\,=\,a} = -D\left(\frac{\partial^3 w}{\partial x^3} + (2 - \nu)\frac{\partial^3 w}{\partial x\,\partial y^2}\right)_{x\,=\,a} = 0. \quad (1.23b)$$

It should be remembered that the twisting moment is the stress resultant from the horizontal shear stresses, and that this twisting moment has now been replaced by statically equivalent vertical forces. This replacement can be done, as pointed out by Thomson and Tait,[7] without altering the bending of the plate as a whole or producing any disturbances in its stress or strain except very near the boundary.

1.10   CONCENTRATED FORCES AT THE CORNERS

If the process of reducing the twisting moment stress resultant and the transverse shear stress resultant to an equivalent transverse shear stress resultant, described in this section and illustrated in Fig. 1.11, is investigated at a corner, a net corner force is obtained. Consider the corner at x = a and y = b in Fig. 1.13.

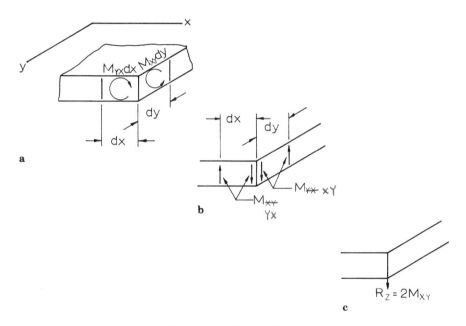

Fig. 1.13.   Corner force

A concentrated force M_{xy} is left over on each face, which is a consequence of the replacement, described earlier, of the twisting moment by vertical forces. The "leftover" forces at the corner of each face, Fig. 1.13b, result in a net corner reaction of $2M_{xy}$ as shown in Fig. 1.13c. Thus, there is required a force of $2M_{xy}$ at each corner of a rectangular plate, supported around the edges in some manner and under the action of a transverse load, to prevent middle surface deflection at the corners.

The corners of a rectangular plate under the action of a uniformly distributed transverse load tend to rise. This action is prevented by the concentrated reactions at the corners.

Again consider the problem of a square plate under the action of a uniformly distributed transverse load. The distribution of the equivalent transverse shear stress resultant along an edge determined by the approximate theory presented here may be compared to the findings of a more exact theory presented by Kromm,[9] for which the effect of transverse shear strain is considered. The comparison is shown in Fig. 1.14. The dashed line shows the results obtained from the approxi-

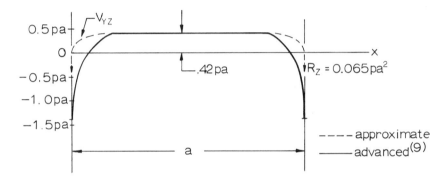

Fig. 1.14.   Comparison of transverse shear force distribution from approximate theory and more advanced theory

mate theory. Note the presence of concentrated corner forces. The solid line shows the continuous distribution obtained from the more exact theory. Note the absence of concentrated corner reactions here. This comparison substantiates the comment made earlier regarding the unreliability of Assumption 2 at certain edges such as corners of plates; however, it also substantiates the reliability of the approximate theory along the remaining portion of the edge.

1.11   SUMMARY

The development of the governing partial differential equation defining small lateral deflections of the middle surface of thin plates, as well as the development of the companion relationships, is based on certain assumptions adopted because of prior knowledge about the behavior of beams. Because of the limitation of small deflections,

1. the middle surface is assumed to remain unstrained. Because of the limitation of thinness,

2. normals to the middle surface before deformation remain normal after deformation, and

3. normal stresses in the direction transverse to the plate are neglected. A fourth assumption based on the material properties is that

4. the material is homogeneous, isotropic, continuous, and linearly elastic.

Governing Equations

$$\nabla^4 w = \frac{\partial^4 w}{\partial x^4} + 2 \frac{\partial^4 w}{\partial x^2 \partial y^2} + \frac{\partial^4 w}{\partial y^2} = \frac{p}{D}$$

or

$$\left\{ \begin{array}{l} \nabla^2 M = \dfrac{\partial^2 M}{\partial x^2} = \dfrac{\partial^2 M}{\partial y^2} = p \\[3mm] \nabla^2 w = \dfrac{\partial^2 w}{\partial x^2} + \dfrac{\partial^2 w}{\partial y^2} = \dfrac{M}{D} \end{array} \right] \quad \text{where} \quad M = -\frac{M_x + M_y}{1 + \nu}$$

Displacements

$$u = -\mathfrak{z} \frac{\partial w}{\partial x}$$

$$v = -\mathfrak{z} \frac{\partial w}{\partial y}$$

Strains

$$\epsilon_x = -\mathfrak{z} \frac{\partial^2 w}{\partial x^2}$$

$$\epsilon_y = -\mathfrak{z} \frac{\partial^2 w}{\partial y^2}$$

$$\epsilon_z = -\frac{\nu}{1 - \nu} (\epsilon_x + \epsilon_y)$$

$$\gamma_{xy} = -2\mathfrak{z} \frac{\partial^2 w}{\partial x \partial y}$$

It should be remembered that the effect of transverse shear strains, γ_{yz} and γ_{xz}, is neglected for the determination of the displacements; thus, their expressions are not required in the development.

Stresses

$$\sigma_x = -\vartheta\,\frac{E}{1-\nu^2}\left(\frac{\partial^2 w}{\partial x^2} + \nu\,\frac{\partial^2 w}{\partial y^2}\right)$$

$$\sigma_y = -\vartheta\,\frac{E}{1-\nu^2}\left(\nu\,\frac{\partial^2 w}{\partial x^2} + \frac{\partial^2 w}{\partial y^2}\right)$$

$$\sigma_z = \text{negligible}$$

$$\tau_{xy} = -\vartheta\,\frac{E}{1+\nu}\,\frac{\partial^2 w}{\partial x\,\partial y}$$

Stress Resultants

$$M_x = -D\left(\frac{\partial^2 w}{\partial x^2} + \nu\,\frac{\partial^2 w}{\partial y^2}\right)$$

$$M_y = -D\left(\nu\,\frac{\partial^2 w}{\partial x^2} + \frac{\partial^2 w}{\partial y^2}\right)$$

$$M_{xy} = M_{yx} = D(1-\nu)\,\frac{\partial^2 w}{\partial x\,\partial y}$$

$$Q_x = Q_y = Q_{xy} = Q_{yx} = 0$$

$$Q_{xz} = -D\,\frac{\partial}{\partial x}\,\nabla^2 w = -D\left(\frac{\partial^3 w}{\partial x^3} + \frac{\partial^3 w}{\partial x\,\partial y^2}\right)$$

$$Q_{yz} = -D\,\frac{\partial}{\partial y}\,\nabla^2 w = -D\left(\frac{\partial^3 w}{\partial x^2\,\partial y} + \frac{\partial^3 w}{\partial y^3}\right)$$

TABLE 1
Boundary Conditions
(Refer to Fig. 1.1)

	Simply Supported	Clamped	Free
On an edge parallel to the y axis at x = a	$w = 0$ $\dfrac{\partial^2 w}{\partial x^2} = 0$	$w = 0$ $\dfrac{\partial w}{\partial x} = 0$	$\dfrac{\partial^2 w}{\partial x^2} + \nu\,\dfrac{\partial^2 w}{\partial y^2} = 0$ $\dfrac{\partial^3 w}{\partial x^3} + (2-\nu)\,\dfrac{\partial^3 w}{\partial x\,\partial y^2} = 0$
On an edge parallel to the x axis at y = b	$w = 0$ $\dfrac{\partial^2 w}{\partial y^2} = 0$	$w = 0$ $\dfrac{\partial w}{\partial y} = 0$	$\nu\,\dfrac{\partial^2 w}{\partial x^2} + \dfrac{\partial^2 w}{\partial y^2} = 0$ $\dfrac{\partial^3 w}{\partial y^3} + (2-\nu)\,\dfrac{\partial^3 w}{\partial x^2\,\partial y} = 0$

Problems

1. What are the assumptions upon which the governing equation and its companion relationships are based? Compare these assumptions to those used in simple beam theory.

2. Which assumption(s) is violated if
   a. a plate has a maximum deflection equivalent to 10 times its thickness?
   b. the thickness of a plate is approximately 20 percent of its width?

3. The expression for the lateral deflection of the simply supported plate in Fig. 1.15 is the double Fourier sine series

$$w(x, y) = \sum_{m=1}^{\infty} \sum_{n=1}^{\infty} A_{mn} \sin \frac{m\pi x}{a} \sin \frac{n\pi y}{b}.$$

a. Verify that this expression satisfies the boundary conditions for a simply supported plate.

b. Determine expressions for the displacements, strains, stresses, and stress resultants.

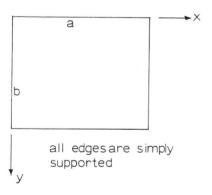

Fig. 1.15

4. For the plate in Problem 3, determine the expressions for the maximum values of the stresses σ_x and τ_{xy}, and indicate where these maximum values occur.

5. For the plate in Problem 3, determine the expressions for the maximum values of the stress resultants M_x and Q_{xz}, and indicate where these maximum values occur.

6. A rectangular plate has two sides simply supported, one side clamped, and one side free, as shown in Fig. 1.16. An approximate

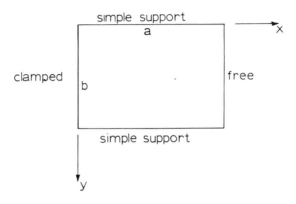

Fig. 1.16

expression for the lateral deflection is

$$w = A \left(\frac{x}{a} \right)^2 \sin \frac{\pi y}{b}$$

where A is a constant.

a. Determine which boundary conditions are satisfied and which are not.

b. Determine approximate expressions for the stress resultants M_x and Q_{xz}, and show on a sketch how each varies with respect to x and y.

c. Determine approximate expressions for the stresses σ_x, σ_y, and τ_{xy}, and show on a sketch how each varies with respect to x and y.

d. Explain why the stresses or stress resultants developed from an approximate expression for the lateral deflection might be highly inaccurate.

7. A rectangular plate has one corner cut off so that an oblique edge is formed, as shown in Fig. 1.17. Determine the expressions for the stress resultants $M_x{}'$, $M_{x'y'}{}'$, and $Q_{x'z}{}'$ on the oblique edge in terms of the angle θ and the coordinates x and y.

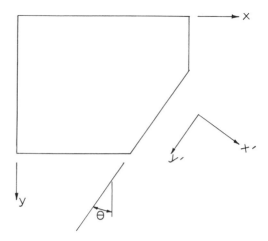

Fig. 1.17

Chapter 2
BASIC RELATIONSHIPS
FOR CIRCULAR PLATES

2.1 INTRODUCTION

We may use the procedure presented in Chapter 1 for rectangular plates to derive the basic relationships for the lateral deflections of circular plates. However, instead of using this procedure a second time, we shall obtain the basic relationships for circular plates from a transformation which relates polar coordinates to Cartesian coordinates. This transformation will be applied to the previously derived equations for rectangular plates in Cartesian coordinates to obtain the similar equations for circular plates in polar coordinates.

The assumptions and the discussion of the assumptions for rectangular plates presented in Section 1.1 apply as well to circular plates. The reader should have a thorough understanding of the contents of Chapter 1 before undertaking the study of this chapter.

2.2 COORDINATES AND DIMENSIONS

The polar coordinates r and θ lie in the middle surface of the plate before deformation, as shown in Fig. 2.1. The coordinate r represents a linear measure along the radius of the plate, and the coordinate θ represents an angular measure in the plane of the middle surface of the plate.

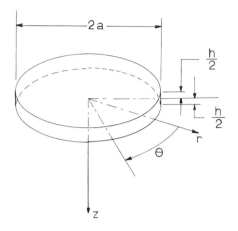

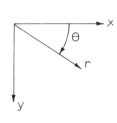

Fig. 2.2. Cartesian
and polar coordinate
systems

Fig. 2.1. Circular plate coordinates
and dimensions

The z axis completes the coordinate system. The deflection of the middle surface is w; the distance of a point in the plate measured from the middle surface is given by the coordinate ζ; and the thickness of the plate is h. The symbols z, w, ζ, and h are defined here exactly as in Section 1.2.

2.3 TRANSFORMATION FROM CARTESIAN TO POLAR COORDINATES

The equations for rectangular plates in Chapter 1 can be modified to apply to circular plates by application of the transformation which relates Cartesian coordinates to polar coordinates. From Fig. 2.2 the equations describing this transformation are

$$x = r \cos \theta \qquad (2.1a)$$

$$y = r \sin \theta. \qquad (2.1b)$$

The following expressions are obtained directly from Eqs. (2.1).

$$r^2 = x^2 + y^2 \qquad (2.2a)$$

$$\theta = \arctan \frac{y}{x} \qquad (2.2b)$$

$$\frac{\partial r}{\partial x} = \cos \theta \tag{2.2c}$$

$$\frac{\partial r}{\partial y} = \sin \theta \tag{2.2d}$$

$$\frac{\partial \theta}{\partial x} = -\frac{1}{r} \sin \theta \tag{2.2e}$$

$$\frac{\partial \theta}{\partial y} = \frac{1}{r} \cos \theta \tag{2.2f}$$

We can now determine the partial derivatives of the deflection surface w with respect to the Cartesian coordinates x and y in terms of the polar coordinates r and θ by using the chain rule of partial differentiation and Eqs. (2.2c)–(2.2f).

$$\frac{\partial w}{\partial x} = \frac{\partial w}{\partial r}\frac{\partial r}{\partial x} + \frac{\partial w}{\partial \theta}\frac{\partial \theta}{\partial x} = \frac{\partial w}{\partial r}\cos \theta - \frac{\partial w}{\partial \theta}\frac{1}{r}\sin \theta \tag{2.3a}$$

$$\frac{\partial w}{\partial y} = \frac{\partial w}{\partial r}\frac{\partial r}{\partial y} + \frac{\partial w}{\partial \theta}\frac{\partial \theta}{\partial y} = \frac{\partial w}{\partial r}\sin \theta + \frac{\partial w}{\partial \theta}\frac{1}{r}\cos \theta \tag{2.3b}$$

$$\frac{\partial^2 w}{\partial x^2} = \frac{\partial}{\partial r}\left(\frac{\partial w}{\partial x}\right)\frac{\partial r}{\partial x} + \frac{\partial}{\partial \theta}\left(\frac{\partial w}{\partial x}\right)\frac{\partial \theta}{\partial x}$$

$$= \left(\frac{\partial^2 w}{\partial r^2}\cos \theta - \frac{\partial^2 w}{\partial r \partial \theta}\frac{1}{r}\sin \theta + \frac{\partial w}{\partial \theta}\frac{1}{r^2}\sin \theta\right)\cos \theta$$

$$- \left(\frac{\partial^2 w}{\partial r \partial \theta}\cos \theta - \frac{\partial w}{\partial r}\sin \theta - \frac{\partial^2 w}{\partial \theta^2}\frac{1}{r}\sin \theta - \frac{\partial w}{\partial \theta}\frac{1}{r}\cos \theta\right)\frac{1}{r}\sin \theta \tag{2.3c}$$

$$\frac{\partial^2 w}{\partial y^2} = \frac{\partial}{\partial r}\left(\frac{\partial w}{\partial y}\right)\frac{\partial r}{\partial y} + \frac{\partial}{\partial \theta}\left(\frac{\partial w}{\partial y}\right)\frac{\partial \theta}{\partial y}$$

$$= \left(\frac{\partial^2 w}{\partial r^2}\sin \theta + \frac{\partial^2 w}{\partial r \partial \theta}\frac{1}{r}\cos \theta - \frac{\partial w}{\partial \theta}\frac{1}{r^2}\cos \theta\right)\sin \theta$$

$$+ \left(\frac{\partial^2 w}{\partial r \partial \theta}\sin \theta + \frac{\partial w}{\partial r}\cos \theta + \frac{\partial^2 w}{\partial \theta^2}\frac{1}{r}\cos \theta - \frac{\partial w}{\partial \theta}\frac{1}{r}\sin \theta\right)\frac{1}{r}\cos \theta \tag{2.3d}$$

$$\frac{\partial^2 w}{\partial x \partial y} = \frac{\partial}{\partial r}\left(\frac{\partial w}{\partial y}\right)\frac{\partial r}{\partial x} + \frac{\partial}{\partial \theta}\left(\frac{\partial w}{\partial y}\right)\frac{\partial \theta}{\partial x}$$

$$= \left(\frac{\partial^2 w}{\partial r^2}\sin \theta - \frac{\partial w}{\partial \theta}\frac{1}{r^2}\cos \theta + \frac{\partial^2 w}{\partial r \partial \theta}\frac{1}{r}\cos \theta\right)\cos \theta$$

$$- \left(\frac{\partial w}{\partial r}\cos \theta + \frac{\partial^2 w}{\partial r \partial \theta}\sin \theta + \frac{\partial^2 w}{\partial \theta^2}\frac{1}{r}\cos \theta - \frac{\partial w}{\partial \theta}\frac{1}{r}\sin \theta\right)\frac{1}{r}\sin \theta \tag{2.3e}$$

2.4 DISPLACEMENTS

The components of linear displacement associated with the polar co-ordinates r and θ are u_r and u_θ. The displacement u_r is positive in the direction of the positive radial coordinate r, and the displacement u_θ is positive in the direction of increasing θ perpendicular to the radial coordinate as shown in Fig. 2.3.

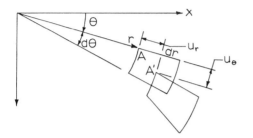
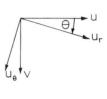

a. Differential element in polar co-ordinates before and after defor-mation illustrating displacement components, u_r and u_θ of point A

b. Positive directions of linear dis-placements in polar and Cartesian coordinates

Fig. 2.3

Let's express the displacements u and v, given by Eqs. (1.2), in terms of the polar coordinates r and θ by using the chain rule of partial dif-ferentiation and Eqs. (2.2c)–(2.2e).

$$u = -\gamma\frac{\partial w}{\partial x} = -\gamma\left(\frac{\partial w}{\partial r}\frac{\partial r}{\partial x} + \frac{\partial w}{\partial\theta}\frac{\partial\theta}{\partial x}\right)$$

$$= -\gamma\left(\frac{\partial w}{\partial r}\cos\theta - \frac{\partial w}{\partial\theta}\frac{1}{r}\sin\theta\right) \qquad (2.4a)$$

$$v = -\gamma\frac{\partial w}{\partial y} = -\gamma\left(\frac{\partial w}{\partial r}\frac{\partial r}{\partial y} + \frac{\partial w}{\partial\theta}\frac{\partial\theta}{\partial y}\right)$$

$$= -\gamma\left(\frac{\partial w}{\partial r}\sin\theta + \frac{\partial w}{\partial\theta}\frac{1}{r}\cos\theta\right) \qquad (2.4b)$$

From inspection of Fig. 2.3, we see that the radial component of dis-placement u_r becomes identical to the displacement u if we allow the radius r to coincide with the x axis by letting $\theta = 0$. That is

$$u_r = u\bigg|_{\theta\,=\,0} = -\gamma\frac{\partial w}{\partial r}\bigg|_{\theta\,=\,0}.$$

Since the position of the x axis is arbitrary, this expression for u_r applies to any radial line; thus, the general expression for the radial displacement is

$$u_r = -\mathfrak{z}\,\frac{\partial w}{\partial r}. \tag{2.5a}$$

Similarly, we can obtain the general expression for the displacement component u_θ by setting $\theta = 0$ in the expression for v.

$$u_\theta = -\mathfrak{z}\,\frac{1}{r}\,\frac{\partial w}{\partial \theta} \tag{2.5b}$$

2.5 STRAINS

The strain components ϵ_x, ϵ_y, and γ_{xy} can be expressed in terms of the polar coordinates r and θ by using Eqs. (2.3c)–(2.3e).

$$\epsilon_x = -\mathfrak{z}\,\frac{\partial^2 w}{\partial x^2} = -\mathfrak{z}\,\cos\theta\left(\frac{\partial^2 w}{\partial r^2}\cos\theta - \frac{\partial^2 w}{\partial r\partial\theta}\frac{1}{r}\sin\theta + \frac{\partial w}{\partial\theta}\frac{1}{r^2}\sin\theta\right)$$

$$+\mathfrak{z}\,\frac{1}{r}\sin\theta\left(\frac{\partial^2 w}{\partial r\partial\theta}\cos\theta - \frac{\partial w}{\partial r}\sin\theta - \frac{\partial^2 w}{\partial\theta^2}\frac{1}{r}\sin\theta - \frac{\partial w}{\partial\theta}\frac{1}{r}\cos\theta\right) \tag{2.6a}$$

$$\epsilon_y = -\mathfrak{z}\,\frac{\partial^2 w}{\partial y^2} = -\mathfrak{z}\,\sin\theta\left(\frac{\partial^2 w}{\partial r^2}\sin\theta + \frac{\partial^2 w}{\partial r\partial\theta}\frac{1}{r}\cos\theta - \frac{\partial w}{\partial\theta}\frac{1}{r^2}\cos\theta\right)$$

$$-\mathfrak{z}\,\frac{1}{r}\cos\theta\left(\frac{\partial^2 w}{\partial r\partial\theta}\sin\theta + \frac{\partial w}{\partial r}\cos\theta + \frac{\partial^2 w}{\partial\theta^2}\frac{1}{r}\cos\theta - \frac{\partial w}{\partial\theta}\frac{1}{r}\sin\theta\right) \tag{2.6b}$$

$$\gamma_{xy} = -2\,\mathfrak{z}\,\frac{\partial^2 w}{\partial x\partial y} = -2\,\mathfrak{z}\,\cos\theta\left(\frac{\partial^2 w}{\partial r^2}\sin\theta - \frac{\partial w}{\partial\theta}\frac{1}{r^2}\cos\theta + \frac{\partial^2 w}{\partial r\partial\theta}\frac{1}{r}\cos\theta\right)$$

$$+2\,\mathfrak{z}\,\frac{1}{r}\sin\theta\left(\frac{\partial w}{\partial r}\cos\theta + \frac{\partial^2 w}{\partial r\partial\theta}\sin\theta + \frac{\partial^2 w}{\partial\theta^2}\frac{1}{r}\cos\theta - \frac{\partial w}{\partial\theta}\frac{1}{r}\sin\theta\right) \tag{2.6c}$$

We may determine expressions for the components of strain in polar coordinates in the same manner that we determined the components of displacement. The normal strains ϵ_r and ϵ_θ and the shear strain $\gamma_{r\theta}$ associated with the polar coordinates are identical to the strains ϵ_x, ϵ_y, and γ_{xy} respectively, if we again allow the radius r to coincide with the x axis by letting $\theta = 0$.

$$\epsilon_r = \epsilon_x \Big|_{\theta = 0} = -\mathfrak{z}\, \frac{\partial^2 w}{\partial r^2}\Big|_{\theta = 0}$$

$$\epsilon_\theta = \epsilon_y \Big|_{\theta = 0} = -\mathfrak{z}\left(\frac{1}{r}\frac{\partial w}{\partial r} + \frac{1}{r^2}\frac{\partial^2 w}{\partial \theta^2}\right)_{\theta = 0}$$

$$\gamma_{r\theta} = \gamma_{xy}\Big|_{\theta = 0} = 2\,\mathfrak{z}\left(\frac{1}{r^2}\frac{\partial w}{\partial \theta} - \frac{1}{r}\frac{\partial^2 w}{\partial r\partial \theta}\right)_{\theta = 0}$$

Since the position of the x axis is arbitrary, these expressions apply to any radial line; thus

$$\epsilon_r = -\mathfrak{z}\, \frac{\partial^2 w}{\partial r^2} \tag{2.7a}$$

$$\epsilon_\theta = -\mathfrak{z}\left(\frac{1}{r}\frac{\partial w}{\partial r} + \frac{1}{r^2}\frac{\partial^2 w}{\partial \theta^2}\right) \tag{2.7b}$$

$$\gamma_{r\theta} = 2\,\mathfrak{z}\left(\frac{1}{r^2}\frac{\partial w}{\partial \theta} - \frac{1}{r}\frac{\partial^2 w}{\partial r\partial \theta}\right). \tag{2.7c}$$

The shear strain components γ_{rz} and $\gamma_{\theta z}$ play the same role here as do γ_{xz} and γ_{yz} for the rectangular plates in Chapter 1. That is, they are considered to have no effect on the magnitudes of the displacements; thus, their expressions are not required in the present development.

2.6 STRESSES

The relationships between stresses and strains in cylindrical coordinates (r, θ, z) for a homogeneous, isotropic, continuous, and linearly elastic material are as follows.

$$\epsilon_r = \frac{1}{E}\left(\sigma_r - \nu\sigma_\theta - \nu\sigma_z\right) \tag{2.8a}$$

$$\epsilon_\theta = \frac{1}{E}\left(-\nu\sigma_r + \sigma_\theta - \nu\sigma_z\right) \tag{2.8b}$$

$$\epsilon_z = \frac{1}{E}\left(-\nu\sigma_r - \nu\sigma_\theta + \sigma_z\right) \tag{2.8c}$$

$$\gamma_{r\theta} = \frac{1}{E} 2(1 + \nu)\, \tau_{r\theta} \qquad (2.8d)$$

As we mentioned in Section 2.5, expressions for the two components of transverse shear strain, $\gamma_{\theta z}$ and γ_{rz}, are not required in the present development. Since the normal stress σ_z is considered small compared with other stresses, according to Assumption 3 in Section 1.1, we can write Eqs. (2.8) in the following form.

$$\sigma_r = \frac{E}{1 - \nu^2} (\epsilon_r + \nu\epsilon_\theta) \qquad (2.9a)$$

$$\sigma_\theta = \frac{E}{1 - \nu^2} (\nu\epsilon_r + \epsilon_\theta) \qquad (2.9b)$$

$$\sigma_z = \text{negligible, or} \quad \sigma_r + \sigma_\theta = -\frac{E}{\nu} \epsilon_z \qquad (2.9c)$$

$$\tau_{r\theta} = \frac{E}{2(1 + \nu)} \gamma_{r\theta} \qquad (2.9d)$$

The stresses of Eqs. (2.9) can now be expressed in terms of the deflection of the middle surface by referring to Eqs. (2.7).

$$\sigma_r = -\frac{\mathfrak{z} E}{1 - \nu^2} \left(\frac{\partial^2 w}{\partial r^2} + \frac{\nu}{r} \frac{\partial w}{\partial r} + \frac{\nu}{r^2} \frac{\partial^2 w}{\partial \theta^2} \right) \qquad (2.10a)$$

$$\sigma_\theta = -\frac{\mathfrak{z} E}{1 - \nu^2} \left(\nu \frac{\partial^2 w}{\partial r^2} + \frac{1}{r} \frac{\partial w}{\partial r} + \frac{1}{r^2} \frac{\partial^2 w}{\partial \theta^2} \right) \qquad (2.10b)$$

$$\tau_{r\theta} = \frac{\mathfrak{z} E}{1 + \nu} \left(\frac{1}{r^2} \frac{\partial w}{\partial \theta} - \frac{1}{r} \frac{\partial^2 w}{\partial r \partial \theta} \right) \qquad (2.10c)$$

2.7 STRESS RESULTANTS

The positive state of stress resultants for a circular plate is shown on the plate element in Fig. 2.4. Figure 2.4 is similar to Fig. 1.9, in which the stress resultants for a rectangular plate are shown.

The stress resultants for a rectangular plate may be expressed in terms of the polar coordinates r and θ if we substitute Eqs. (2.3) into (1.11). From a comparison of Figs. 1.9 and 2.4, we may again argue that

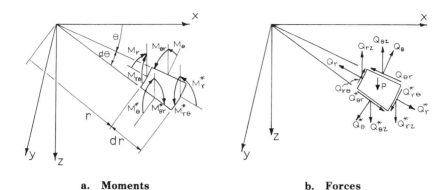

a. Moments   **b. Forces**

Fig. 2.4. Differential plate element (middle surface) with stress resultants

the stress resultants for a circular plate are identical to those for a rectangular plate if we allow the radius r to coincide with the x axis by letting $\theta = 0$:

$$\mathbf{M}_r = \mathbf{M}_x \Big|_{\theta = 0}$$

$$\mathbf{M}_\theta = \mathbf{M}_y \Big|_{\theta = 0}$$

$$\mathbf{M}_{r\theta} = \mathbf{M}_{\theta r} = \mathbf{M}_{xy} \Big|_{\theta = 0}$$

$$\mathbf{Q}_{rz} = \mathbf{Q}_{xz} \Big|_{\theta = 0}$$

$$\mathbf{Q}_{\theta z} = \mathbf{Q}_{yz} \Big|_{\theta = 0}$$

$$\mathbf{Q}_r = \mathbf{Q}_\theta = \mathbf{Q}_{r\theta} = \mathbf{Q}_{\theta r} = 0.$$

Since the position of the x axis is arbitrary, these expressions apply to any radial line, and the general expressions for the stress resultants for a circular plate are

$$\mathbf{M}_r = -\mathbf{D}\left(\frac{\partial^2 \mathbf{w}}{\partial \mathbf{r}^2} + \frac{\nu}{\mathbf{r}}\frac{\partial \mathbf{w}}{\partial \mathbf{r}} + \frac{\nu}{\mathbf{r}^2}\frac{\partial^2 \mathbf{w}}{\partial \theta^2}\right) \qquad (2.11\text{a})$$

$$\mathbf{M}_\theta = -\mathbf{D}\left(\frac{1}{\mathbf{r}}\frac{\partial \mathbf{w}}{\partial \mathbf{r}} + \frac{1}{\mathbf{r}^2}\frac{\partial^2 \mathbf{w}}{\partial \theta^2} + \nu\frac{\partial^2 \mathbf{w}}{\partial \mathbf{r}^2}\right) \qquad (2.11\text{b})$$

$$M_{r\theta} = M_{\theta r} = D(1 - \nu)\left(\frac{1}{r}\frac{\partial^2 w}{\partial r \partial \theta} - \frac{1}{r^2}\frac{\partial w}{\partial \theta}\right) \qquad (2.11c)$$

$$Q_{rz} = -D\frac{\partial}{\partial r}(\nabla^2 w) \qquad (2.11d)$$

$$Q_{\theta z} = -D\frac{1}{r}\frac{\partial}{\partial \theta}(\nabla^2 w) \qquad (2.11e)$$

$$Q_r = Q_\theta = Q_{r\theta} = Q_{\theta r} = 0. \qquad (2.11f)$$

2.8 GOVERNING EQUATION

We should recall that the harmonic operator in Cartesian coordinates is

$$\nabla^2 = \frac{\partial^2}{\partial x^2} + \frac{\partial^2}{\partial y^2}. \qquad (2.12)$$

The harmonic operator can be expressed in terms of polar coordinates by substitution of Eqs. (2.3c) and (2.3d) into (2.12).

$$\nabla^2 = \frac{\partial^2}{\partial r^2} + \frac{1}{r}\frac{\partial}{\partial r} + \frac{1}{r^2}\frac{\partial^2}{\partial \theta^2} \qquad (2.13)$$

Thus, the partial differential equation defining the deflection of the middle surface of the plate [Eq. (1.15c)] becomes

$$\nabla^2\nabla^2 w = \frac{p}{D}$$

$$\left(\frac{\partial^2}{\partial r^2} + \frac{1}{r}\frac{\partial}{\partial r} + \frac{1}{r^2}\frac{\partial^2}{\partial \theta^2}\right)\left(\frac{\partial^2}{\partial r^2} + \frac{1}{r}\frac{\partial}{\partial r} + \frac{1}{r^2}\frac{\partial^2}{\partial \theta^2}\right)w = \frac{p}{D} \qquad (2.14a)$$

or

$$\frac{\partial^4 w}{\partial r^4} + \frac{2}{r}\frac{\partial^3 w}{\partial r^3} - \frac{1}{r^2}\frac{\partial^2 w}{\partial r^2} + \frac{1}{r^3}\frac{\partial w}{\partial r} + \frac{2}{r^2}\frac{\partial^4 w}{\partial r^2 \partial \theta^2}$$

$$- \frac{2}{r^3}\frac{\partial^3 w}{\partial \theta^2 \partial r} + \frac{4}{r^4}\frac{\partial^2 w}{\partial \theta^2} + \frac{1}{r^4}\frac{\partial^4 w}{\partial \theta^4} = \frac{p}{D}. \qquad (2.14b)$$

2.9   BOUNDARY CONDITIONS

A complete solution of the governing partial differential equation for a circular plate depends upon the conditions of the plate at the boundaries. These conditions must be expressed in terms of the deflection of the middle surface. Three types of boundaries are considered: simply supported, clamped, and free.

Simply Supported Edge Conditions. A plate boundary that is prevented from deflecting but free to rotate about a line along the boundary edge, such as a hinge, is defined as a simply supported edge. The conditions on a simply supported edge at $r = a$ and any θ are

$$w\Big|_{r\,=\,a} = 0$$

$$M_r\Big|_{r\,=\,a} = 0 = -D\left(\frac{\partial^2 w}{\partial r^2} + \frac{\nu}{r}\frac{\partial w}{\partial r} + \frac{\nu}{r^2}\frac{\partial^2 w}{\partial \theta^2}\right)_{r\,=\,a}.$$

Since the change of w with respect to the θ coordinate vanishes along this edge, these conditions become

$$w\Big|_{r\,=\,a} = 0 \tag{2.15a}$$

$$\left(\frac{\partial^2 w}{\partial r^2} + \frac{\nu}{r}\frac{\partial w}{\partial r}\right)_{r\,=\,a} = 0. \tag{2.15b}$$

The boundary conditions on a simply supported edge along a radius at $\theta = \beta$ and any r are

$$w\Big|_{\theta\,=\,\beta} = 0$$

$$M_\theta\Big|_{\theta\,=\,\beta} = 0 = -D\left(\frac{1}{r}\frac{\partial w}{\partial r} + \frac{1}{r^2}\frac{\partial^2 w}{\partial \theta^2} + \nu\frac{\partial^2 w}{\partial r^2}\right)_{\theta\,=\,\beta}.$$

Since the change of w with respect to the r coordinate vanishes along this edge, these conditions become

$$w\Big|_{\theta\,=\,\beta} = 0 \tag{2.16a}$$

$$\left.\frac{\partial^2 w}{\partial \theta^2}\right|_{\theta = \beta} = 0. \tag{2.16b}$$

Clamped Edge Conditions. If a plate boundary is clamped, the deflection and slope of the middle surface must be zero at the boundary. The boundary conditions on a clamped edge at $r = a$ and any θ are

$$\left. w \right|_{r = a} = 0 \tag{2.17a}$$

$$\left.\frac{\partial w}{\partial r}\right|_{r = a} = 0. \tag{2.17b}$$

The boundary conditions on a clamped edge along a radius at $\theta = \beta$ and any r are

$$\left. w \right|_{\theta = \beta} = 0 \tag{2.18a}$$

$$\left.\frac{\partial w}{\partial \theta}\right|_{\theta = \beta} = 0. \tag{2.18b}$$

Free Edge Conditions. In the most general case, a twisting moment, a bending moment, and a transverse shear force act on an edge of a plate, as shown in Fig. 2.4. An edge on which all three of these stress resultants vanish is defined as a free edge. Since two boundary conditions at each edge are sufficient for a complete solution of the governing equation, the twisting moment and the transverse shear force are combined into one boundary condition. This condition is referred to as the Kirchhoff shear force, as discussed in Section 1.9. Thus, the Kirchhoff shear force and the bending moment must vanish on a free edge. These two boundary conditions on the free edge at $r = a$ and any θ, as shown in Fig. 2.5, are

$$\left. M_r \right|_{r = a} = 0 = -D\left(\frac{\partial^2 w}{\partial r^2} + \frac{\nu}{r}\frac{\partial w}{\partial r} + \frac{\nu}{r^2}\frac{\partial^2 w}{\partial \theta^2}\right)_{r = a} \tag{2.19a}$$

$$\left. V_{rz} \right|_{r = a} = 0 = \left(Q_{rz} - \frac{1}{r}\frac{\partial M_{r\theta}}{\partial \theta}\right)_{r = a}$$

$$= -D\left[\frac{\partial}{\partial r}(\nabla^2 w) + \frac{1 - \nu}{r}\frac{\partial}{\partial \theta}\left(\frac{1}{r}\frac{\partial^2 w}{\partial r \partial \theta} - \frac{1}{r^2}\frac{\partial w}{\partial \theta}\right)\right]_{r = a}. \tag{2.19b}$$

The boundary conditions on the free edge at $\theta = \beta$ and any r, as shown in Fig. 2.5, are

$$\mathbf{M}_\theta\Big|_{\theta = \beta} = 0 = -D\left(\frac{1}{r}\frac{\partial w}{\partial r} + \frac{1}{r^2}\frac{\partial^2 w}{\partial \theta^2} + \nu\frac{\partial^2 w}{\partial r^2}\right)_{\theta = \beta} \qquad (2.20a)$$

$$\mathbf{V}_{\theta z}\Big|_{\theta = \beta} = 0 = \left(\mathbf{Q}_{\theta z} - \frac{\partial \mathbf{M}_{r\theta}}{\partial r}\right)_{\theta = \beta}$$

$$= -D\left[\frac{1}{r}\frac{\partial}{\partial \theta}(\nabla^2 w) + (1 - \nu)\frac{\partial}{\partial r}\left(\frac{1}{r}\frac{\partial^2 w}{\partial r\partial\theta} - \frac{1}{r^2}\frac{\partial w}{\partial \theta}\right)\right]_{\theta = \beta}. \qquad (2.20b)$$

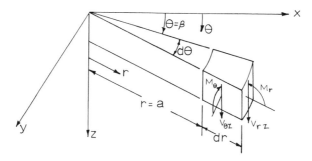

Fig. 2.5.   Positive state of reduced stress resultants for a free edge

2.10   SUMMARY

The derivations of the partial differential equation that defines the small lateral deflections of thin circular plates and the companion relationships are based on the transformation relating Cartesian coordinates to polar coordinates. This transformation is applied to the equations previously developed for rectangular plates to obtain the corresponding equations for circular plates.

Governing Equation

$$\nabla^4 w = \nabla^2\nabla^2 w = \left(\frac{\partial^2}{\partial r^2} + \frac{1}{r}\frac{\partial}{\partial r} + \frac{1}{r^2}\frac{\partial^2}{\partial \theta^2}\right)\left(\frac{\partial^2}{\partial r^2} + \frac{1}{r}\frac{\partial}{\partial r} + \frac{1}{r^2}\frac{\partial^2}{\partial \theta^2}\right)w = \frac{p}{D}$$

Displacements

$$u_r = -z\frac{1}{r}\frac{\partial w}{\partial r}$$

$$u_\theta = -3 \frac{1}{r} \frac{\partial w}{\partial \theta}$$

Strains

$$\epsilon_r = -3 \frac{\partial^2 w}{\partial r^2}$$

$$\epsilon_\theta = -3 \left(\frac{1}{r} \frac{\partial w}{\partial r} + \frac{1}{r^2} \frac{\partial^2 w}{\partial \theta^2} \right)$$

$$\epsilon_z = -\frac{\nu}{1-\nu} (\epsilon_r + \epsilon_\theta)$$

$$\gamma_{r\theta} = 2\,3 \left(\frac{1}{r^2} \frac{\partial w}{\partial \theta} - \frac{1}{r} \frac{\partial^2 w}{\partial r \partial \theta} \right)$$

Remember that the effect of transverse shear strains, γ_{rz} and $\gamma_{\theta z}$, is neglected for the determination of the displacements; thus, their expressions are not required in the development.

Stresses

$$\sigma_r = -3 \frac{E}{1-\nu^2} \left(\frac{\partial^2 w}{\partial r^2} + \frac{\nu}{r} \frac{\partial w}{\partial r} + \frac{\nu}{r^2} \frac{\partial^2 w}{\partial \theta^2} \right)$$

$$\sigma_\theta = -3 \frac{E}{1-\nu^2} \left(\nu \frac{\partial^2 w}{\partial r^2} + \frac{1}{r} \frac{\partial w}{\partial r} + \frac{1}{r^2} \frac{\partial^2 w}{\partial \theta^2} \right)$$

$$\sigma_z = \text{negligible}$$

$$\tau_{r\theta} = \frac{E}{1-\nu} \left(\frac{1}{r^2} \frac{\partial w}{\partial \theta} - \frac{1}{r} \frac{\partial^2 w}{\partial r \partial \theta} \right)$$

Stress Resultants

$$M_r = -D \left(\frac{\partial^2 w}{\partial r^2} + \frac{\nu}{r} \frac{\partial w}{\partial r} + \frac{\nu}{r^2} \frac{\partial^2 w}{\partial \theta^2} \right)$$

$$M_\theta = -D \left(\frac{1}{r} \frac{\partial w}{\partial r} + \frac{1}{r^2} \frac{\partial^2 w}{\partial \theta^2} + \nu \frac{\partial^2 w}{\partial r^2} \right)$$

$$M_{r\theta} = D(1-\nu) \left(\frac{1}{r} \frac{\partial^2 w}{\partial r \partial \theta} - \frac{1}{r^2} \frac{\partial w}{\partial \theta} \right)$$

$$Q_{rz} = -D \frac{\partial}{\partial r} (\nabla^2 w)$$

$$Q_{\theta z} = -D \frac{1}{r} \frac{\partial}{\partial \theta} (\nabla^2 w)$$

$$Q_r = Q_\theta = Q_{r\theta} = Q_{\theta r} = 0$$

TABLE 2
Boundary Conditions

	Simply Supported	Clamped	Free
On an edge at r = a and any θ	$\frac{\partial^2 w}{\partial r^2} + \frac{\nu}{r} \frac{\partial w}{\partial r} = 0$ w = 0	w = 0 $\frac{\partial w}{\partial r} = 0$	$M_r = 0$ $Q_{rz} - \frac{1}{r} \frac{\partial M_{r\theta}}{\partial \theta} = 0$
On an edge along a radius at θ = β and any r	w = 0 $\frac{\partial^2 w}{\partial \theta^2} = 0$	w = 0 $\frac{\partial w}{\partial \theta} = 0$	$M_\theta = 0$ $Q_{\theta z} - \frac{\partial M_{r\theta}}{\partial r} = 0$

Problems

8. The expression for the lateral deflection of the simply supported circular plate in Fig. 2.6 is

$$w = \frac{a^4}{32 D} \left(\frac{3 + \nu}{1 + \nu} - \frac{p_0}{2 D} \right) - \frac{a^2}{32 D} \left(\frac{3 + \nu}{1 + \nu} \right) r^2 + \frac{p_0 r^4}{64 D}.$$

a. Verify that this expression satisfies the boundary conditions for a simply supported circular plate.

b. Determine expressions for the displacements, strains, stresses, and stress resultants.

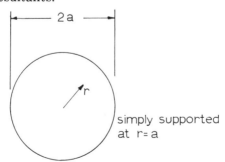

simply supported
at r= a

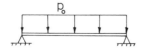

Fig. 2.6

9. For the plate in Problem 8, determine the expressions for the maximum values of σ_r and σ_θ, and indicate where these maximum values occur.

10. For the plate in Problem 8, determine the expressions for the maximum values of the stress resultants M_r, M_θ, and Q_{rz}, and indicate where these maximum values occur.

11. The circular plate in Fig. 2.7 has its interior edge clamped and its exterior edge free. An approximate expression for the lateral deflection is $w = A(r-a)^2$, where A is a constant.

   a. Determine which boundary conditions are satisfied and which are not.

   b. Determine an approximate expression for each stress resultant.

   c. Determine an approximate expression for each stress.

   d. Explain why the stresses or stress resultants developed from an approximate expression for the lateral deflection might be highly inaccurate.

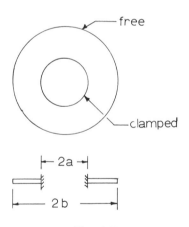

Fig. 2.7

12. Derive expressions for the stress resultants M_r, M_θ, and $M_{r\theta}$ in terms of the lateral deflection, using the procedure presented in Section 1.6 for rectangular plates.

13. Draw a free body diagram of a differential element in polar coordinates, and

   a. derive the equilibrium equations for a circular plate analogous to Eqs. (1.12a), (1.12b), and (1.12c) for a rectangular plate;

b. derive the equilibrium equation for a circular plate analogous to Eq. (1.13) for a rectangular plate.

c. From the results of Problem 12 and part b of this problem, derive the governing equation for a circular plate analogous to Eq. (1.15) for a rectangular plate.

CLASSICAL SOLUTIONS

Chapter 3

CLASSICAL SOLUTIONS FOR
THE LATERAL DEFLECTIONS OF
RECTANGULAR PLATES

3.1 THE NAVIER SOLUTION FOR THE LATERAL DEFLECTIONS OF SIMPLY SUPPORTED RECTANGULAR PLATES

As we proceed, we shall find that the solutions to lateral deflections of thin plates often become very involved. Many times the deflections, stress resultants, and stresses are obtained in terms of very complicated infinite series. In the past, summing these series presented formidable tasks. However, with the increasing availability of digital computers these problems become much more readily solved. We will find in Chapters 7 and 8 that certain numerical techniques have been developed specifically for the digital computer. However, before we examine some of the more recent developments in thin plate theory, it is important that we develop a general background in the field of thin plates.

The authors do not intend to minimize the value of the more classical solutions presented in this chapter. On the contrary, many problems have been, and can still be, more easily solved by the classical techniques which we will discuss.

All solutions for the lateral deflections of thin plates must satisfy the governing differential equation given by Eq. (1.15), and the appropriate plate boundary conditions developed in the preceding chapters. The type of boundary conditions to which the plate is subjected plays a major part in the solution technique applicable to each particular problem. The boundary conditions which lead to the simplest plate problems occur when all four edges of the plate are simply supported, Fig. 3.1.

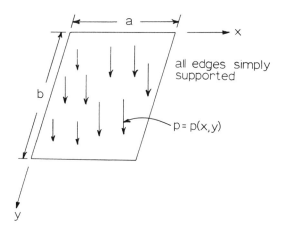

Fig. 3.1. Simply supported rectangular plate with an arbitrary transverse load

The boundary conditions for a simply supported, or hinged, rectangular plate are given by Eqs. (1.17) and (1.18) in the form

$$w\Big|_{x\,=\,0} = w\Big|_{x\,=\,a} = 0$$

$$\frac{\partial^2 w}{\partial x^2}\Big|_{x\,=\,0} = \frac{\partial^2 w}{\partial x^2}\Big|_{x\,=\,a} = 0$$

and

$$w\Big|_{y\,=\,0} = w\Big|_{y\,=\,b} = 0$$

$$\frac{\partial^2 w}{\partial y^2}\Big|_{y\,=\,0} = \frac{\partial^2 w}{\partial y^2}\Big|_{y\,=\,b} = 0.$$

The first solution to be developed for the lateral deflections of a simply supported rectangular plate under an arbitrary transverse loading, p(x,y), is attributed to Navier.[1,11] Navier noted that the boundary conditions are automatically satisfied by choosing a deflection function in the form

$$w(x,y) = \sum_{m=1}^{\infty} \sum_{n=1}^{\infty} W_{mn} \sin \frac{m\pi x}{a} \sin \frac{n\pi y}{b}. \tag{3.1}$$

From the theory of Fourier series we know that Eq. (3.1) can be made to represent any piecewise continuous function over the range of $0 \le x \le a$ and $0 \le y \le b$. Consequently, Eq. (3.1) will define the deflection $w(x,y)$ if we can determine the proper coefficients W_{mn}. Thus, we complete the problem by determining the coefficients W_{mn} in such a manner that Eq. (3.1) satisfies the partial differential equation $\nabla^4 w = p(x,y)/D$. Substitution of Eq. (3.1) into the partial differential equation yields

$$\sum_{m=1}^{\infty} \sum_{n=1}^{\infty} W_{mn} \left[\left(\frac{m\pi}{a} \right)^2 + \left(\frac{n\pi}{b} \right)^2 \right]^2 \sin \frac{m\pi x}{a} \sin \frac{n\pi y}{b} = \frac{p(x,y)}{D}. \tag{3.2}$$

We can determine the coefficients W_{mn} from Eq. (3.2) by first expressing the known load distribution in the form of a double Fourier sine series[12]

$$p(x,y) = \sum_{m=1}^{\infty} \sum_{n=1}^{\infty} A_{mn} \sin \frac{m\pi x}{a} \sin \frac{n\pi y}{b} \tag{3.3}$$

in which

$$A_{mn} = \frac{4}{ab} \int_0^b \int_0^a p(x,y) \sin \frac{m\pi x}{a} \sin \frac{n\pi y}{b} \, dxdy. \tag{3.4}$$

Substitution of Eq. (3.3) into (3.2) yields

$$\sum_{m=1}^{\infty} \sum_{n=1}^{\infty} W_{mn} \pi^4 \left[\left(\frac{m}{a} \right)^2 + \left(\frac{n}{b} \right)^2 \right]^2 \sin \frac{m\pi x}{a} \sin \frac{n\pi y}{b}$$

$$= \sum_{m=1}^{\infty} \sum_{n=1}^{\infty} \frac{A_{mn}}{D} \sin \frac{m\pi x}{a} \sin \frac{n\pi y}{b}$$

or

$$\sum_{m=1}^{\infty} \sum_{n=1}^{\infty} \left\{ W_{mn} \pi^4 \left[\left(\frac{m}{a} \right)^2 + \left(\frac{n}{b} \right)^2 \right]^2 - \frac{A_{mn}}{D} \right\}$$

$$\times \sin \frac{m\pi x}{a} \sin \frac{n\pi y}{b} = 0. \tag{3.5}$$

Equation (3.5) must be satisfied for all values of x and y. Consequently, the total coefficient must equal zero for all values of m and n, i.e.

$$W_{mn}\pi^4\left[\left(\frac{m}{a}\right)^2 + \left(\frac{n}{b}\right)^2\right]^2 - \frac{A_{mn}}{D} = 0$$

or

$$W_{mn} = \frac{A_{mn}}{\pi^4 D\left[\left(\frac{m}{a}\right)^2 + \left(\frac{n}{b}\right)^2\right]^2} . \tag{3.6}$$

Substitution of Eq. (3.4) into (3.6) yields

$$W_{mn} = \frac{4}{ab\pi^4 D\left[\left(\frac{m}{a}\right)^2 + \left(\frac{n}{b}\right)^2\right]^2}$$

$$\times \int_0^b \int_0^a p(x,y) \sin\frac{m\pi x}{a} \sin\frac{n\pi y}{b} \, dxdy. \tag{3.7}$$

Hence, the double Fourier sine series given in Eq. (3.1) defines the lateral deflection of a simply supported thin rectangular plate under the action of an arbitrary transverse loading if W_{mn} is determined from Eq. (3.7).

The infinite series given by Eq. (3.1) converges rapidly in many cases. Only a few terms need be included in the solution for such cases. As an example, this solution applied to a square plate under the action of a uniform load yields results with error of approximately 2.5 percent if only the first term in the series is included.[1] However, in general, convergence must be checked by examining the individual effects of several terms. We should note that after the integration specified in Eq. (3.4) is completed, the infinite series given in Eq. (3.1) can be programmed on a digital computer with relative ease.

Stresses can be computed by substituting Eq. (3.1) into Eqs. (1.10a), (1.10b), and (1.1k). However, the series obtained for the stresses generally converge much more slowly than the series given in Eq. (3.1).

Another important illustration of the Navier solution is a simply supported rectangular plate with a uniform transverse load distributed over the small area, as shown in Fig. 3.2. The solution of this problem

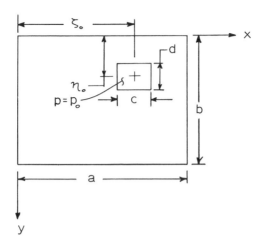

Fig. 3.2.   Simply supported plate with uniform loading on small segment

will prove useful in leading to the idea of influence coefficients, as we shall see shortly. For this example, Eq. (3.7) becomes

$$W_{mn} = \cfrac{4}{ab\pi^4 D \left[\left(\cfrac{m}{a}\right)^2 + \left(\cfrac{n}{b}\right)^2\right]^2}$$

$$\times \int_{\eta_0 - (d/2)}^{\eta_0 + (d/2)} \int_{\zeta_0 - (c/2)}^{\zeta_0 + (c/2)} p_0 \sin\frac{m\pi x}{a} \sin\frac{n\pi y}{b}\, dx\, dy. \qquad (3.8)$$

After integration, Eq. (3.8) gives

$$W_{mn} = \cfrac{16 p_0}{mn\pi^6 D \left[\left(\cfrac{m}{a}\right)^2 + \left(\cfrac{n}{b}\right)^2\right]^2} \sin\frac{m\pi\zeta_0}{a} \sin\frac{m\pi c}{2a} \sin\frac{n\pi\eta_0}{b} \sin\frac{n\pi d}{2b}. \qquad (3.9)$$

Substitution of Eq. (3.9) into (3.1) yields the following expression for the lateral deflections.

$$w(x,y) = \sum_{m=1}^{\infty}\sum_{n=1}^{\infty}\left\{\cfrac{16 p_0}{mn\pi^6 D \left[\left(\cfrac{m}{a}\right)^2 + \left(\cfrac{n}{b}\right)^2\right]^2} \sin\frac{m\pi\zeta_0}{a} \sin\frac{m\pi c}{2a}\right.$$

$$\left. \times \sin\frac{n\pi\eta_0}{b} \sin\frac{n\pi d}{2b}\right\}\sin\frac{m\pi x}{a} \sin\frac{n\pi y}{b} \qquad (3.10)$$

The actual numerical value of the deflection is determined at any location of interest, (x,y), by substituting the appropriate values of p_0, x, y, ζ_0, and η_0 into Eq. (3.10). We can extend this solution to the problem of a concentrated load at (ζ_0, η_0) by first defining the total load on the cd area as P. Then

$$P = p_0 \, cd \qquad\qquad (3.11a)$$

or

$$p_0 = \frac{P}{cd}. \qquad\qquad (3.11b)$$

Substitution of Eq. (3.11b) into (3.9) results in

$$W_{mn} = \frac{16P}{mncd\pi^6 D \left[\left(\dfrac{m}{a}\right)^2 + \left(\dfrac{n}{b}\right)^2 \right]^2}$$
$$\times \sin \frac{m\pi\zeta_0}{a} \sin \frac{m\pi c}{2a} \sin \frac{n\pi\eta_0}{b} \sin \frac{n\pi d}{2b}. \qquad (3.12)$$

If c and d are made to approach zero while we let p_0 approach infinity such that the product defined by Eq. (3.11a) remains constant, then in the limit the load becomes a concentrated load of magnitude P acting at the point (ζ_0, η_0), and Eq. (3.12) becomes

$$W_{mn} = \lim_{c\&d\to 0} \left\{ \frac{16P}{mncd\pi^6 D \left[\left(\dfrac{m}{a}\right)^2 + \left(\dfrac{n}{b}\right)^2 \right]^2} \right.$$
$$\times \sin \frac{m\pi\zeta_0}{a} \sin \frac{m\pi c}{2a} \sin \frac{n\pi\eta_0}{b} \sin \frac{n\pi d}{2b} \left. \right\}. \qquad (3.13)$$

This expression is evaluated by examining the individual limits

$$\lim_{c\to 0} \frac{\sin \dfrac{m\pi c}{2a}}{c} = \frac{0}{0}$$

and

$$\lim_{d\to 0} \frac{\sin \dfrac{n\pi d}{2b}}{d} = \frac{0}{0}.$$

Note that both of these limits are undefined in the present form. However, we evaluate them by applying L'Hospital's rule,[13]

$$\lim_{c \to 0} \frac{\sin \dfrac{m\pi c}{2a}}{c} = \lim_{c \to 0} \frac{m\pi}{2a} \cos \frac{m\pi c}{2a} = \frac{m\pi}{2a} \qquad (3.14a)$$

and

$$\lim_{d \to 0} \frac{\sin \dfrac{n\pi d}{2b}}{d} = \lim_{d \to 0} \frac{n\pi}{2b} \cos \frac{n\pi d}{2b} = \frac{n\pi}{2b}. \qquad (3.14b)$$

When we utilize the limits given in Eqs. (3.14a) and (3.14b), Eq. (3.13) becomes

$$W_{mn} = \frac{4P}{ab\pi^4 D \left[\left(\dfrac{m}{a}\right)^2 + \left(\dfrac{n}{b}\right)^2 \right]^2} \sin \frac{m\pi \zeta_0}{a} \sin \frac{n\pi \eta_0}{b}. \qquad (3.15)$$

Thus, the lateral deflections at any point (x,y) due to a concentrated load of magnitude P applied at the point (ζ_0, η_0) is

$$w(x,y) = \sum_{m=1}^{\infty} \sum_{n=1}^{\infty} \frac{4P}{ab\pi^4 D \left[\left(\dfrac{m}{a}\right)^2 + \left(\dfrac{n}{b}\right)^2 \right]^2} \sin \frac{m\pi \zeta_0}{a} \sin \frac{n\pi \eta_0}{b}$$
$$\times \sin \frac{m\pi x}{a} \sin \frac{n\pi y}{b}. \qquad (3.16)$$

This result is further extended to the concept of influence coefficients by considering Eq. (3.16) for the special case when P = 1.

$$w(x,y) = \sum_{m=1}^{\infty} \sum_{n=1}^{\infty} \frac{4}{ab\pi^4 D \left[\left(\dfrac{m}{a}\right)^2 + \left(\dfrac{n}{b}\right)^2 \right]^2} \sin \frac{m\pi \zeta_0}{a} \sin \frac{n\pi \eta_0}{b}$$
$$\times \sin \frac{m\pi x}{a} \sin \frac{n\pi y}{b} = K(x,y; \zeta_0, \eta_0) \qquad (3.17a)$$

The function $K(x,y; \zeta_0, \eta_0)$, which describes the lateral deflections due to a unit load at (ζ_0, η_0), is defined as an influence coefficient. If we have a concentrated load P which acts at (ζ_0, η_0), and denoted by $P(\zeta_0, \eta_0)$, then the lateral deflections resulting from this load are described by

$$w(x,y) = P(\zeta_0, \eta_0) \ K(x,y; \zeta_0, \eta_0). \tag{3.17b}$$

If n concentrated loads exist on the plate, we can apply the principle of superposition to obtain the resultant deflection, since deflections are linearly related to applied loads. Thus, we have

$$w(x,y) = w_1(x,y) + w_2(x,y) + \ldots \ldots w_j(x,y)$$
$$= P_1(\zeta_1, \eta_1) \ K(x,y; \zeta_1, \eta_1) + P_2(\zeta_2, \eta_2) \ K(x,y; \zeta_2, \eta_2)$$
$$+ \ldots \ldots + P_j(\zeta_j, \eta_j) \ K(x,y; \zeta_j, \eta_j)$$
$$= \sum_{i=1}^{j} P_i(\zeta_i, \eta_i) \ K(x,y; \zeta_i, \eta_i). \tag{3.17c}$$

Finally, we treat a continuous load with the same technique by visualizing the continuous load as an infinite number of concentrated loads, each of magnitude $p(\zeta, \eta)dxdy$. The resulting continuous infinite summation simply becomes an integration over the area to which the continuous load is applied. Thus, we write

$$w(x,y) = \int_A p(\zeta, \eta) \ K(x,y; \zeta, \eta) \ d\zeta d\eta \tag{3.17d}$$

in which the area of integration, A, corresponds to the plate area over which the load $p(\zeta, \eta)$ exists.

3.2 THE LEVY SOLUTION FOR THE LATERAL DEFLECTIONS OF RECTANGULAR PLATES

The Navier solution is very straightforward. It would be very much in the spirit of the Constitution if we could apply it to all thin plate problems regardless of race, creed, religion, shape, or boundary conditions. Unfortunately, it applies only to the limited category of simply supported rectangular plates.

A more general technique which yields the lateral deflections of plates with boundary conditions other than simply supported was developed by Levy.[1,3] The Levy solution yields the lateral deflections of rectangular plates with two opposite edges simply supported, and with arbitrary boundary conditions imposed on the remaining two sides as shown in Fig. 3.3.

In Section 3.3 we will find that the Levy solution, in combination with the principle of superposition, can be utilized in the deflection analyses of rectangular plates with arbitrary boundary conditions imposed on all four sides.

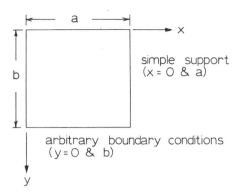

Fig. 3.3. Plate boundary conditions corresponding to the Levy solution

The two basic criteria which the solution must again satisfy are (1) the boundary conditions, and (2) the partial differential equation $\nabla^4 w = p/D$. We begin by assuming the total solution to consist of a homogeneous part and a particular part

$$w = w_h + w_p \tag{3.18}$$

in which w_h and w_p represent the homogeneous and particular solutions respectively. This assumption leads to the equation

$$\nabla^4 w_h = 0 \tag{3.19}$$

from which the homogeneous solution is determined, and to the equation

$$\nabla^4 w_p = \frac{p(x,y)}{D} \tag{3.20}$$

from which the particular solution is determined.

Since Eq. (3.19) is independent of the loading, a single homogeneous solution can be developed for all rectangular plates that have two opposite sides simply supported. The particular solution must be determined for each individual loading condition, $p(x,y)$.

The homogeneous solution is assumed in the form

$$w_h = \sum_{m=1}^{\infty} f_m(y) \sin \frac{m\pi x}{a} \tag{3.21}$$

in which $f_m(y)$ is an arbitrary function of y. Equation (3.21) automatically satisfies the simply supported boundary conditions along the edges $x = 0$ and $x = a$. The problem is solved by first determining the function $f_m(y)$ in such a manner that w_h satisfies the two simply supported boundary conditions and the differential equation given by Eq. (3.19). The solution is completed by requiring the total solution, $w = w_h + w_p$, to satisfy the boundary conditions on the two arbitrary sides. Substitution of the assumed solution given by Eq. (3.21) into (3.19) yields

$$\sum_{m=1}^{\infty} \left[\left(\frac{m\pi}{a} \right)^4 f_m - 2 \left(\frac{m\pi}{a} \right)^2 \frac{d^2 f_m}{dy^2} + \frac{d^4 f_m}{dy^4} \right] \sin \frac{m\pi x}{a} = 0. \qquad (3.22)$$

Derivatives of f_m are total derivatives instead of partial derivatives since f_m is a function of y only. Equation (3.22) must be satisfied for all values of x. Hence

$$\left(\frac{m\pi}{a} \right)^4 f_m - 2 \left(\frac{m\pi}{a} \right)^2 \frac{d^2 f_m}{dy^2} + \frac{d^4 f_m}{dy^4} = 0. \qquad (3.23)$$

Since Eq. (3.23) is an ordinary linear differential equation with constant coefficients, it obviously has a solution of the type

$$f_m = E_m e^{\lambda_m y} \qquad (3.24)$$

in which E_m and λ_m are constants. If in the previous sentence the word "obviously" caused you a twinge of pain, then you are advised to refer to a text on ordinary differential equations for a quick review. If we substitute Eq. (3.24) into (3.23), the following characteristic equation is obtained.

$$\left(\frac{m\pi}{a} \right)^4 - 2 \left(\frac{m\pi}{a} \right)^2 \lambda_m^2 + \lambda_m^4 = 0$$

or

$$\left[\lambda_m^2 - \left(\frac{m\pi}{a} \right)^2 \right] \left[\lambda_m^2 - \left(\frac{m\pi}{a} \right)^2 \right] = 0$$

from which the repeated roots

$$\lambda_m = + \frac{m\pi}{a}, + \frac{m\pi}{a}, - \frac{m\pi}{a}, - \frac{m\pi}{a}$$

are obtained.

At this point in the derivation, the constant E_m has not been specified, and thus it can have a different value corresponding to each root λ_m. The most general solution for $f_m(y)$ must contain the sum of all possible solutions. Consequently

$$f_m(y) = A'_m e^{m\pi y/a} + B'_m e^{-m\pi y/a} + C'_m y e^{m\pi y/a} + D'_m y e^{-m\pi y/a} \qquad (3.25)$$

in which A'_m, B'_m, C'_m, and D'_m are arbitrary constants. Use of the trigonometric identities

$$\sinh \frac{m\pi y}{a} = \frac{1}{2}(e^{m\pi y/a} - e^{-m\pi y/a})$$

$$\cosh \frac{m\pi y}{a} = \frac{1}{2}(e^{m\pi y/a} + e^{-m\pi y/a})$$

allows us to rewrite Eq. (3.25) in the form

$$f_m(y) = A_m \sinh \frac{m\pi y}{a} + B_m \cosh \frac{m\pi y}{a}$$
$$+ C_m y \sinh \frac{m\pi y}{a} + D_m y \cosh \frac{m\pi y}{a} \qquad (3.26)$$

where the primed and unprimed constants are related as follows.

$$A_m = A_m' - B_m' \, ; \, \big| \, B_m = A_m' + B_m'$$

$$C_m = C_m' - D_m' \, ; \, \big| \, D_m = C_m' + D_m'$$

Substitution of Eq. (3.26) into (3.21) gives the homogeneous solution in terms of the constants A_m, B_m, C_m, and D_m.

$$w_h = \sum_{m=1}^{\infty} \left[A_m \sinh \frac{m\pi y}{a} + B_m \cosh \frac{m\pi y}{a} \right.$$
$$\left. + C_m y \sinh \frac{m\pi y}{a} + D_m y \cosh \frac{m\pi y}{a} \right] \sin \frac{m\pi x}{a} \qquad (3.27)$$

We will later determine the four constants by requiring that the solution, $w = w_h + w_p$, satisfy the four boundary conditions on the two arbitrary edges which have yet to be satisfied. However, before we can determine these constants, the particular solution must be obtained.

There are many ways to determine particular solutions to differential equations. Sometimes the determination of a particular solution to a differential equation is very difficult, and any technique which allows the engineer to obtain such a solution is not only acceptable but marvelous. A frequently used technique is described in the following paragraph.

The particular solution may be determined by first developing the load, p(x,y), in a single Fourier sine series[4]

$$p(x,y) = \sum_{m=1}^{\infty} p_m(y) \sin \frac{m\pi x}{a} \qquad (3.28)$$

in which

$$p_m(y) = \frac{2}{a} \int_0^a p(x,y) \sin \frac{m\pi x}{a} \, dx.$$

Next, we assume w_p in the form

$$w_p = \sum_{m=1}^{\infty} k_m(y) \sin \frac{m\pi x}{a} \qquad (3.29)$$

in which $k_m(y)$ is an arbitrary function of y. Note that w_p automatically satisfies the boundary conditions of $w_p = 0$ and $\partial^2 w_p / \partial x^2 = 0$ along the edges at $x = 0$ and $x = a$. We previously determined w_h in a form that satisfied these boundary conditions, and hence the complete solution, $w = w_h + w_p$, also satisfies these conditions.

Substitution of Eqs. (3.28) and (3.29) into Eq. (3.20) yields

$$\sum_{m=1}^{\infty} \left[\frac{d^4 k_m}{dy^4} - 2\left(\frac{m\pi}{a}\right)^2 \frac{d^2 k_m}{dy^2} + \left(\frac{m\pi}{a}\right)^4 k_m \right] \sin \frac{m\pi x}{a}$$

$$= \sum_{m=1}^{\infty} \frac{p_m(y)}{D} \sin \frac{m\pi x}{a}. \qquad (3.30)$$

Since Eq. (3.30) must be satisfied independent of x

$$\frac{d^4 k_m}{dy^4} - 2\left(\frac{m\pi}{a}\right)^2 \frac{d^2 k_m}{dy^2} + \left(\frac{m\pi}{a}\right)^4 k_m = \frac{p_m}{D}. \qquad (3.31)$$

The solution of w_p is completed by determining a particular solution, k_m, to the ordinary differential equation given by Eq. (3.31). The final solution is determined by requiring that $w = w_h + w_p$ satisfy the four expressions defining the boundary conditions along the edges at y = 0 and y = b. These four boundary conditions yield four equations from which the four constants A_m, B_m, C_m, and D_m are determined.

The solution of the similar problem with arbitrary boundary conditions along the sides x = 0 and x = a is obtained by interchanging x and a with y and b respectively.

Uniformly Loaded Plate with All Edges Simply Supported. The Levy solution can be easily illustrated if we consider a simply supported plate with a uniform load. Based on Section 3.1, it is evident that this problem can also be solved by the Navier solution. However, the resulting solution of this particular problem, in the form of a Levy solution, is of further use to us in the following section on superposition; consequently, it is to our advantage to consider it as an example at this point in our studies.

For this example, we assume that the edges x = 0 and x = a are simply supported; and we initially treat the edges y = 0 and y = b as arbitrary. The homogeneous solution is given by Eq. (3.27). We next must obtain the particular solution. Thus, in accordance with Eq. (3.28) the uniform load p_0 is developed in the form

$$p(x,y) = \sum_{m=1}^{\infty} p_m(y) \sin \frac{m\pi x}{a}.$$

The coefficient $p_m(y)$ is of the form

$$p_m(y) = \frac{2}{a} \int_0^a p_0 \sin \frac{m\pi x}{a} dx \stackrel{p_0}{=} \left(\frac{2}{a}\right)\left(\frac{a}{m\pi}\right)\left(-\cos \frac{m\pi x}{a}\right)_0^a$$

$$= \begin{cases} \dfrac{4p_0}{m\pi} & \text{for } m = 1, 3, 5, \ldots \\ 0 & \text{for } m = 2, 4, 6, \ldots \end{cases}$$

Thus, Eq. (3.31) becomes

$$\frac{d^4 k_m}{dy^4} - 2\left(\frac{m\pi}{a}\right)^2 \frac{d^2 k_m}{dy^2} + \left(\frac{m\pi}{a}\right)^4 k_m = \frac{4p_0}{m\pi D}, \quad m = 1, 3, 5, \ldots \quad (3.32)$$

We are very fortunate in this problem since careful inspection of Eq. (3.32) reveals the particular solution of this ordinary differential equation to be

$$k_m(y) = \frac{4p_0 a^4}{m^5 \pi^5 D}, \quad m = 1, 3, 5, \ldots \quad (3.33)$$

Substitution of Eq. (3.33) into (3.29) gives the particular solution of the governing partial differential equation as

$$w_p = \sum_{m=1,3,5,\ldots}^{\infty} \frac{4p_0 a^4}{m^5 \pi^5 D} \sin \frac{m\pi x}{a}. \tag{3.34}$$

We obtain the total solution by adding the particular and the homogeneous solutions, i.e. Eqs. (3.34) and (3.27).

$$w = \sum_{m=1}^{\infty} \left(A_m \sinh \frac{m\pi y}{a} + B_m \cosh \frac{m\pi y}{a} \right.$$

$$+ C_m y \sinh \frac{m\pi y}{a} + D_m y \cosh \frac{m\pi y}{a} \right) \sin \frac{m\pi x}{a}$$

$$+ \sum_{m=1,3,5,\ldots}^{\infty} \frac{4p_0 a^4}{m^5 \pi^5 D} \sin \frac{m\pi x}{a} \tag{3.35}$$

We now must consider the boundary conditions of the two boundaries which we designated as arbitrary, $y = 0$ and $y = b$.

The remainder of the problem is simplified by changing the location of the coordinate system to take advantage of the existing symmetry. Consider the coordinate system in Fig. 3.4.

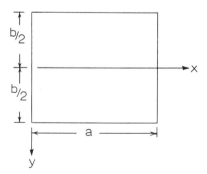

Fig. 3.4.   Revised plate coordinate system

Since the location of the x axis has not yet entered into the solution, Eq. (3.35) is the applicable solution with the new coordinate system. The terms $B_m \cosh m\pi y/a$ and $yC_m \sinh m\pi y/a$ are symmetric about the x axis and are called even functions. The terms $A_m \sinh m\pi y/a$ and yD_m $yB_m \cosh m\pi y/a$ are antisymmetric about the x axis and are called odd functions.

We observe that the lateral deflections are symmetric about the x axis for this problem since the loading and boundary conditions are symmetric. Thus, the coefficient A_m and D_m must equal zero. Similarly, we note that the lateral deflections are symmetric about the line x = a/2. Since sin $m\pi x/a$ yields terms symmetric about this line only if m = 1, 3, 5, . . . , both summations in Eq. (3.35) must include only terms with m = 1, 3, 5, ... Equation (3.35) now reduces to

$$w = \sum_{m=1,3,5,...}^{\infty} \left(B_m \cosh \frac{m\pi y}{a} + C_m y \sinh \frac{m\pi y}{a} \right.$$
$$\left. + \frac{4p_0 a^4}{m^5 \pi^5 D} \right) \sin \frac{m\pi x}{a}. \qquad (3.36)$$

Recall that at this point in the solution the governing partial differential equation is satisfied at all points, and the boundary conditions are satisfied along the edges x = 0 and x = a. Thus, the remaining coefficients B_m and C_m must be determined in such a manner that the boundary conditions along the edges at y = 0 and y = b are satisfied.

The symmetry of Eq. (3.36) about the x axis implies that if the boundary conditions are satisfied along either of the edges y = + b/2 or y = − b/2 then the boundary conditions along both of these edges are satisfied. The boundary condition w = 0 at y = b/2 is satisfied if

$$\sum_{m=1,3,5,...} \left(B_m \cosh \frac{m\pi b}{2a} + C_m \frac{b}{2} \sinh \frac{m\pi b}{2a} \right.$$
$$\left. + \frac{4p_0 a^4}{m^5 \pi^5 D} \right) \sin \frac{m\pi x}{a} = 0. \qquad (3.37)$$

Equation (3.37) must apply for all values of x. Thus

$$B_m \cosh \frac{m\pi b}{2a} + C_m \frac{b}{2} \sinh \frac{m\pi b}{2a} + \frac{4p_0 a^4}{m^5 \pi^5 D} = 0. \qquad (3.38)$$

Evaluation of the boundary condition which specifies zero moment requires the expression for $\partial^2 w/\partial y^2$, which is

$$\frac{\partial^2 w}{\partial y^2} = \sum_{m=1,3,5}^{\infty} \left[B_m \left(\frac{m\pi}{a}\right)^2 \cosh \frac{m\pi y}{a} + 2C_m \left(\frac{m\pi}{a}\right) \cosh \frac{m\pi y}{a} \right.$$
$$\left. + C_m \left(\frac{m\pi}{a}\right)^2 y \sinh \frac{m\pi y}{a} \right] \sin \frac{m\pi x}{a}.$$

Consequently, the boundary condition $\partial^2 w / \partial y^2 = 0$ at $y = b/2$ yields

$$\sum_{m=1,3,5}^{\infty} \left[B_m \left(\frac{m\pi}{a} \right) \cosh \frac{m\pi b}{2a} + 2C_m \cosh \frac{m\pi b}{2a} \right.$$
$$\left. + C_m \left(\frac{m\pi}{a} \right) \left(\frac{b}{2} \right) \sinh \frac{m\pi b}{2a} \right] = 0 \; \sin \frac{m\pi x}{a} = 0 \quad (3.39)$$

Since Eq. (3.39) must also be satisfied for all values of x, then

$$\left[B_m \left(\frac{m\pi}{a} \right) + 2C_m \right] \cosh \frac{m\pi b}{2a} + C_m \left(\frac{m\pi b}{2a} \right) \sinh \frac{m\pi b}{2a} = 0. \quad (3.40)$$

The simultaneous solution of Eqs. (3.38) and (3.40) yields

$$B_m = - \frac{4 p_0 a^4 + m\pi p_0 a^3 b \tanh \dfrac{m\pi b}{2a}}{m^5 \pi^5 D \cosh \dfrac{m\pi b}{2a}} \quad (3.41a)$$

$$C_m = \frac{2 p_0 a^3}{m^4 \pi^4 D \cosh \dfrac{m\pi b}{2a}}. \quad (3.41b)$$

Substitution of Eqs. (3.41a) and (3.41b) into Eq. (3.36) yields the final solution:

$$w = \frac{p_0 a^3}{\pi^5 D} \sum_{m=1,3,5,\ldots}^{\infty} \left(\frac{4a + m\pi b \tanh \dfrac{m\pi b}{2a}}{m^5 \cosh \dfrac{m\pi b}{2a}} \cosh \frac{m\pi y}{a} \right.$$
$$\left. + \frac{2\pi y}{m^4 \cosh \dfrac{m\pi b}{2a}} \sinh \frac{m\pi y}{a} + \frac{4a}{m^5} \right) \sin \frac{m\pi x}{a}. \quad (3.42)$$

This is such a messy looking equation that one might be misled into wishing he hadn't bothered with the problem. Actually, closer examination of Eq. (3.42) reveals that the calculation of deflections at specified points simply involves a series of divisions, multiplications, and additions. As we stated at the opening of this chapter, this task is ideal for a digital computer. Often convergence is quite rapid and hand calculations are feasible.

Uniformly Loaded Plate with Three Sides Simply Supported and One Side Clamped.

As a final example of the Levy technique, let's consider a uniformly loaded plate, simply supported on three sides and clamped on the fourth side.

Note that the coordinate system is moved back to a corner of the plate, as shown in Fig. 3.5, since this problem does not have all of the

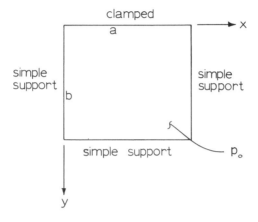

Fig. 3.5. Plate boundary conditions corresponding to example 2

symmetry of the previous problem. The homogeneous and particular solutions are again given by Eqs. (3.27) and (3.34) since the only change is the boundary conditions on the arbitrary sides.

Symmetry again exists in this problem about the line $x = a/2$. Therefore, we again reason that w_h may be summed only for odd integers of m. Hence, the general solution becomes

$$w = \sum_{m=1,3,5,\ldots}^{\infty} \left(A_m \sinh \frac{m\pi y}{a} + B_m \cosh \frac{m\pi y}{a} + C_m y \sinh \frac{m\pi y}{a} \right.$$

$$\left. + D_m y \cosh \frac{m\pi y}{a} + \frac{4p_0 a^4}{m^5 \pi^5 D} \right) \sin \frac{m\pi x}{a}. \tag{3.43}$$

The four constants in Eq. (3.43) are determined from the boundary conditions

$$w \Big|_{y = 0} = 0 \tag{3.44a}$$

$$\frac{\partial w}{\partial y} \Big|_{y = 0} = 0 \tag{3.44b}$$

$$w \Big|_{y \,=\, b} = 0 \qquad\qquad (3.44c)$$

$$\frac{\partial^2 w}{\partial y^2} \bigg|_{y \,=\, b} = 0. \qquad\qquad (3.44d)$$

If we apply the condition given by Eq. (3.44a) to Eq. (3.43), we obtain

$$\sum_{m \,=\, 1,3,5,\ldots}^{\infty} \left(B_m + \frac{4p_0 a^4}{m^5 \pi^5 D} \right) \sin \frac{m\pi x}{a} = 0.$$

This condition must apply for all values of x. Thus

$$B_m = - \frac{4p_0 a^4}{m^5 \pi^5 D}. \qquad\qquad (3.45a)$$

Similarly, the remaining three boundary conditions yield the following expressions.

$$A_m \left(\frac{m\pi}{a} \right) + D_m = 0 \qquad\qquad (3.45b)$$

$$A_m \sinh \frac{m\pi b}{a} + B_m \cosh \frac{m\pi b}{a} + C_m b \sinh \frac{m\pi b}{a}$$
$$+ D_m b \cosh \frac{m\pi b}{a} + \frac{4p_0 a^4}{m^5 \pi^5 D} = 0 \qquad\qquad (3.45c)$$

$$A_m \left(\frac{m\pi}{a} \right)^2 \sinh \frac{m\pi b}{a} + B_m \left(\frac{m\pi}{a} \right)^2 \cosh \frac{m\pi b}{a}$$
$$+ 2C_m \frac{m\pi}{a} \cosh \frac{m\pi b}{a} + C_m \left(\frac{m\pi}{a} \right)^2 b \sinh \frac{m\pi b}{a}$$
$$+ 2D_m \left(\frac{m\pi}{a} \right) \sinh \frac{m\pi b}{a} + D_m \left(\frac{m\pi}{a} \right)^2 b \cosh \frac{m\pi b}{a} = 0. \quad (3.45d)$$

The simultaneous solution of Eqs. (3.45a), (3.45b), (3.45c), and (3.45d) yields the remaining constants:

$$A_m = - \frac{a}{m\pi} D_m$$

$$= \frac{4p_0 a^4}{m^5 \pi^5 D} \left[\frac{2\cosh^2 \dfrac{m\pi b}{a} - 2\cosh \dfrac{m\pi b}{a} - \dfrac{m\pi b}{a} \sinh \dfrac{m\pi b}{a}}{2\cosh \dfrac{m\pi b}{a} \sinh \dfrac{m\pi b}{a} - 2\dfrac{m\pi b}{a}} \right] \quad (3.46a)$$

$$C_m = \frac{4p_0a^4}{m^5\pi^5D}$$

$$\times \left[\frac{2\frac{m\pi}{a}\sinh\frac{m\pi b}{a}\cosh\frac{m\pi b}{a} - \frac{m\pi}{a}\sinh\frac{m\pi b}{a} - \left(\frac{m\pi}{a}\right)^2 b\cosh\frac{m\pi b}{a}}{2\cosh\frac{m\pi b}{a}\sinh\frac{m\pi b}{a} - 2\frac{m\pi b}{a}}\right]. \quad (3.46b)$$

The expression for the total deflection is completed if we substitute these constants into the general solution given by Eq. (3.43). Again the final result appears involved; however, the summation can be easily accomplished with a digital computer.

The application of this technique leads to the expression defining lateral deflections with any other combination of boundary conditions on the two opposite arbitrary sides.

3.3 LATERAL DEFLECTIONS OF RECTANGULAR PLATES WITH ARBITRARY BOUNDARY CONDITIONS BY THE METHOD OF SUPERPOSITION

We can now determine the lateral deflections of a rectangular plate, regardless of the boundary conditions, by combining the Levy solution with the principle of superposition. We begin this technique by reducing the initial problem to several less complex problems, each of which can be solved by the Levy solution. Then the solutions to each of these simple problems are added in such a manner that the governing differential equation and the boundary conditions are satisfied for the initial problem.

Plate with Two Adjacent Sides Simply Supported, One Side Clamped and One Side Free. As an example of the principle of superposition, consider a plate with a uniform load, p_0, as shown in Fig. 3.6.

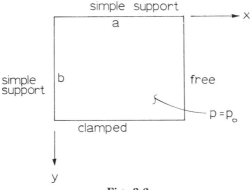

Fig. 3.6

The boundary conditions are

$S.S$

$$w\Big|_{x\,=\,0} = \frac{\partial^2 w}{\partial x^2}\Big|_{x\,=\,0} = 0$$

free B.C's

$$\left(\frac{\partial^2 w}{\partial x^2} + \nu\,\frac{\partial^2 w}{\partial y^2}\right)_{x\,=\,a} = \left(\frac{\partial^3 w}{\partial x^3} + (2-\nu)\,\frac{\partial^3 w}{\partial x \partial y^2}\right)_{x\,=\,a} = 0$$

$S.S.$

$$w\Big|_{y\,=\,0} = \frac{\partial^2 w}{\partial y^2}\Big|_{y\,=\,0} = 0$$

clamped B.C's

$$w\Big|_{y\,=\,b} = \frac{\partial w}{\partial y}\Big|_{y\,=\,b} = 0.$$

Since the governing partial differential equation $\nabla^4 w = p/D$ is linear, the method of superposition applies. The solution to our problem illustrated in Fig. 3.6 may be obtained by the superposition of solutions of each of the three plates illustrated in Fig. 3.7. We observe that the boundary conditions for each of these three plates are such that the Levy solution is applicable in each case. Thus, we have reduced the original complicated problem to three relatively simple problems. We should mention that more than one combination of solutions can be superposed to yield the solution to the original more difficult problem. Thus, considerable care must be taken to determine the most desirable combination because the superposition technique, at its very best, becomes quite involved. The ability to select the most desirable combination will come with experience.

The following governing differential equations apply to the three plates in Fig. 3.7.

$$\nabla^4 w_1 = \frac{p_0}{\textcircled{O}_D} \quad \text{for Plate No. 1} \qquad (3.47a)$$

$$\nabla^4 w_2 = 0 \quad \text{for Plate No. 2} \qquad (3.47b)$$

$$\nabla^4 w_3 = 0 \quad \text{for Plate No. 3} \qquad (3.47c)$$

where w_1, w_2, and w_3 represent the deflections of Plates 1, 2, and 3 respectively.

Upon obtaining the solution for the deflection of each of the three plates in Fig. 3.7 and adding these three solutions to obtain the total

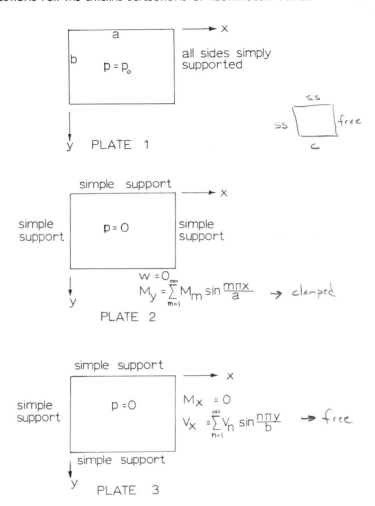

Fig. 3.7.   **Component parts of superposition problem**

solution, we have satisfied every boundary condition except

$$\left. \frac{\partial \mathbf{w}}{\partial \mathbf{y}} \right|_{\mathbf{y} \ = \ \mathbf{b}} = 0$$

and

$$\left. \mathbf{V}_x \right|_{\mathbf{x} \ = \ \mathbf{a}} = 0$$

which are satisfied by the proper choice of the unknown constants M_m and V_n.

We ask the question: "How are the other boundary conditions satisfied?" As an example, let's consider the boundary condition

$$M_x\Big|_{x = 0} = \frac{\partial^2 w}{\partial x^2}\Big|_{x = 0} = 0.$$

The condition of superposition yields

$$w = w_1 + w_2 + w_3.$$

Differentiation gives

$$\frac{\partial^2 w}{\partial x^2} = \frac{\partial^2 w_1}{\partial x^2} + \frac{\partial^2 w_2}{\partial x^2} + \frac{\partial^2 w_3}{\partial x^2}.$$

The solutions for w_1, w_2, and w_3 are based upon the conditions that

$$\frac{\partial^2 w_1}{\partial x^2}\Big|_{x = 0} = \frac{\partial^2 w_2}{\partial x^2}\Big|_{x = 0} = \frac{\partial^2 w_3}{\partial x^2}\Big|_{x = 0} = 0.$$

Therefore

$$\frac{\partial^2 w}{\partial x^2}\Big|_{x = 0} = 0.$$

We make similar arguments to show that the other boundary conditions are satisfied.

The physical interpretation of the functions $\sum_{m=1}^{\infty} M_m \sin \frac{m\pi x}{a}$ and $\sum_{n=1}^{\infty} V_n \sin \frac{n\pi y}{b}$ is relatively simple. The function $\sum_{m=1}^{\infty} M_m \sin \frac{m\pi x}{a}$ represents the necessary bending moment which must exist on the edge $y = b$ to force the slope to be zero. The function $\sum_{n=1}^{\infty} V_n \sin \frac{n\pi y}{b}$ is equal in magnitude and opposite in sign to the sum of the shear along the edge $x = a$ in Plates 1 and 2. Thus, the total shear along this edge of the original plate equals zero.

Now that we have presented an approach to the problem, let's hammer out the solution. We shall take advantage of the fact that the expression defining the lateral deflections of Plate 1 has previously been developed and is given in Eq. (3.42). This equation is adjusted to the coordinate system in Fig. 3.7 by replacing y with y − b/2.

Hence, we can write

$$w_1 = \sum_{m=1,3,5,\ldots}^{\infty} R_m(y) \sin \frac{m\pi x}{a} \tag{3.48}$$

if we define $R_m(y) = \left[\dfrac{4a + m\pi b \tanh \frac{m\pi b}{2a}}{m^5 \cosh \frac{m\pi b}{2a}} \cosh \frac{m\pi\left(y - \frac{b}{2}\right)}{a} + \dfrac{2\pi y}{m^4 \cosh \frac{m\pi b}{2a}} \sinh \frac{m\pi y}{a} + \dfrac{4a}{m^5} \right] \dfrac{p_0 a^3}{\pi^5 D}$

$$R_m(y) = \frac{p_0 a^3}{\pi^5 D} \sum_{m=1,3,5,\ldots}^{\infty} \left[\frac{4a - m\pi p_0 b \tanh \frac{m\pi b}{2a}}{m^5 \cosh \frac{m\pi b}{2a}} \cosh \frac{m\pi\left(y - \frac{b}{2}\right)}{a} \right.$$

$$\left. + \frac{2\pi y}{m^5 \cosh \frac{m\pi b}{2b}} \sinh \frac{m\pi\left(y - \frac{b}{2}\right)}{a} \right] \tag{3.49}$$

The expressions defining the lateral deflections of Plates 2 and 3 contain only a homogeneous part, since p = 0. The lateral deflections of Plate 2 are defined by the expression

$$w_2 = \sum_{m=1}^{\infty} F_m(y) \sin \frac{m\pi x}{a}$$

or

$$w_2 = \sum_{m=1}^{\infty} M_m f_m(y) \sin \frac{m\pi x}{a}. \tag{3.50}$$

The only difference between Eq. (3.50) and the expression for the homogeneous part of the standard Levy solution, Eq. (3.21), is the constant M_m. We will see later that factoring out M_m allows a simplified solution to this problem.

Substitution of Eq. (3.50) into the differential equation given by Eq. (3.47b) results in

$$\sum_{m=1}^{\infty} M_m \left[\left(\frac{m\pi}{a}\right)^4 f_m - 2\left(\frac{m\pi}{a}\right)^2 \frac{d^2 f_m}{dy^2} + \frac{d^4 f_m}{dy^4} \right] \sin \frac{m\pi x}{a} = 0.$$

Since this expression must be satisfied for all values of x

$$\left(\frac{m\pi}{a}\right)^4 f_m - 2\left(\frac{m\pi}{a}\right)^2 \frac{d^2f_m}{dy^2} + \frac{d^4f_m}{dy^4} = 0. \tag{3.51}$$

We see that M_m canceled out, and that Eq. (3.51) is of the same form as Eq. (3.23). Hence

$$f_m(y) = A_{m2} \sinh \frac{m\pi y}{a} + B_{m2} \cosh \frac{m\pi y}{a}$$

$$+ C_{m2}y \sinh \frac{m\pi y}{a} + D_{m2}y \cosh \frac{m\pi y}{a}$$

and

$$w_2 = \sum_{m=1}^{\infty} M_m \left(A_{m2} \sinh \frac{m\pi y}{a} + B_{m2} \cosh \frac{m\pi y}{a} + C_{m2}y \sinh \frac{m\pi y}{a} \right.$$

$$\left. + D_{m2}y \cosh \frac{m\pi y}{a} \right) \sin \frac{m\pi x}{a}. \tag{3.52}$$

The constants A_{m2}, B_{m2}, C_{m2}, and D_{m2} are determined from the conditions

$$\left. \frac{\partial^2 w_2}{\partial y_x^2} \right|_{y=0} = w_2 \Big|_{y=0} = w_2 \Big|_{y=b} = 0 \tag{3.53a}$$

and

$$-D\left(\frac{\partial^2 w_2}{\partial y^2} + \nu \frac{\partial^2 w_2}{\partial x^2} \right)_{y=b} = \sum_{m=1}^{\infty} M_m \sin \frac{m\pi x}{a}. \tag{3.53b}$$

Substitution of the three boundary conditions given in Eq. (3.53a) into (3.52) yields

$$B_{m2} = C_{m2} = 0 \tag{3.54a}$$

and

$$A_{m2} = -D_{m2} b \coth \frac{m\pi b}{a}. \tag{3.54b}$$

Hence, w_2 is written in the simplified form

$$w_2 = \sum_{m=1}^{\infty} M_m \left(-D_{m2} \, b \, \coth \frac{m\pi b}{a} \sinh \frac{m\pi y}{a} \right.$$
$$\left. + D_{m2} \, y \cosh \frac{m\pi y}{a} \right) \sin \frac{m\pi x}{a}. \qquad (3.55)$$

If we substitute Eq. (3.55) into the final boundary condition given by Eq. (3.53b), we obtain

$$-D \sum_{m=1}^{\infty} M_m \, D_{m2} \, S \left(\frac{m\pi b}{a} \right) = \sum_{m=1}^{\infty} M_m \sin \frac{m\pi x}{a} \qquad (3.56)$$

in which

$$S \left(\frac{m\pi b}{a} \right) = \left(\frac{m\pi}{a} \right)^2 \left[\frac{2a}{m\pi} \sinh \frac{m\pi b}{a} + b(1-\nu) \cosh \frac{m\pi b}{a} \right.$$
$$\left. - b(1-\nu) \coth \frac{m\pi b}{a} \sinh \frac{m\pi b}{a} \right].$$

Regrouping the terms in Eq. (3.56) leads to

$$\sum_{m=1}^{\infty} M_m \left[D_{m2} \, S \left(\frac{m\pi b}{a} \right) D + 1 \right] \sin \frac{m\pi x}{a} = 0. \qquad (3.57)$$

Since Eq. (3.57) must be satisfied for all values of x, we must have

$$D_{m2} \, S \left(\frac{m\pi b}{a} \right) D + 1 = 0$$

or

$$D_{m2} = -\frac{1}{D \, S \left(\frac{m\pi b}{a} \right)}. \qquad (3.58)$$

The expression for w_2 becomes

$$w_2 = \sum_{m=1}^{\infty} \frac{M_m}{D \, S \left(\frac{m\pi b}{a} \right)} \left[b \coth \frac{m\pi b}{a} \sinh \frac{m\pi y}{a} - y \cosh \frac{m\pi y}{a} \right] \sin \frac{m\pi x}{a}$$

for clamped - ss B.C's

where the constant M_m is the only unknown.

We can now obtain a solution to Plate 3 in the same manner. If we define the lateral deflection as

$$w_3 = \sum_{n=1}^{\infty} V_n \, g_n(x) \, \sin \frac{n\pi y}{b} \tag{3.59}$$

then we find

$$g_n(x) = A_{n3} \sinh \frac{n\pi x}{b} + B_{n3} \cosh \frac{n\pi x}{b}$$
$$+ C_{n3} \, x \sinh \frac{n\pi x}{b} + D_{n3} x \cosh \frac{n\pi x}{b}. \tag{3.60}$$

We determine the constants A_{n3}, B_{n3}, C_{n3}, and D_{n3} from the conditions

$$\left. w_3 \right|_{x=0} = 0$$

$$\left. \frac{\partial^2 w_3}{\partial x^2} \right|_{x=0} = 0 \qquad\qquad \left.\begin{array}{c} s.s.\ edge \\ B.C.'s \end{array}\right\} \tag{3.61a}$$

and

$$\left(\frac{\partial^2 w_3}{\partial x^2} + \nu \frac{\partial^2 w_3}{\partial y^2} \right)_{x=a} = 0$$

$$\left(\frac{\partial^3 w_3}{\partial x^3} + (2-\nu) \frac{\partial^3 w_3}{\partial x \partial y^2} \right)_{x=a} = \sum_{n=1}^{\infty} V_n \sin \frac{n\pi y}{b} \quad \left.\begin{array}{c} free\ edge \\ B.C.'s \end{array}\right\}. \tag{3.61b}$$

Each of these conditions is satisfied independent of the constant V_n in a manner similar to the solution for Plate 2. Thus, V_n is the only constant which is undetermined at this point in the solution of Plate 3.

The solution to the original plate is now written in the form

$$w = w_1 + w_2 + w_3$$

$$= \sum_{m=1,3,5,\ldots}^{\infty} R_m(y) \sin \frac{m\pi x}{a} + \sum_{m=1}^{\infty} M_m f_m(y) \sin \frac{m\pi x}{a}$$

$$+ \sum_{n=1}^{\infty} V_n \, g_n(x) \, \sin \frac{n\pi y}{b}. \tag{3.62}$$

We should again remember that $R_m(y)$, $f_m(y)$, and $g_n(x)$ are now known functions. The only unknowns are the constants M_m and V_n. Equation (3.62) satisfies the basic differential equation and the boundary conditions with the exception of the following two.

$$\frac{\partial w}{\partial y}\bigg|_{y = b} = 0 \qquad (clamped\ edge) \qquad (3.63a)$$

$$V_x = - D\left[\frac{\partial^3 w}{\partial x^3} + (2-\nu) \frac{\partial^3 w}{\partial x \partial y^2}\right]_{x = a} = 0 \qquad (3.63b)$$

$$free\ edge$$

If we substitute the expression for the total deflection given by Eq. (3.62) into the two boundary conditions given by Eqs. (3.63a) and (3.63b), we obtain two equations that we can solve for the last two unknowns, M_m and V_m. This completes the solution; however, the mathematical juggling involved in this process is more involved than we might expect. Although it is impractical to discuss all of the potential hazards, it is appropriate to discuss at least some of the difficulties most frequently encountered by continuing the solution to this problem.

If we apply the first boundary condition, Eq. (3.63a), to the expression for the deflection given by Eq. (3.62), we obtain

$$\frac{\partial w}{\partial y}\bigg|_{y = b} = \left[\sum_{m=1,3,5,\dots}^{\infty} \frac{dR_m}{dy} \sin \frac{m\pi x}{a}\right.$$

$$+ \sum_{m=1}^{\infty} M_m \frac{df_m}{dy} \sin \frac{m\pi x}{a}$$

$$+ \left.\sum_{n=1}^{\infty} \frac{n\pi}{b} V_n g_n(x) \cos \frac{n\pi y}{b}\right]_{y = b} = 0. \qquad (3.64)$$

To provide a common factor in Eq. (3.64), we express the function $g_n(x)$ in the form

$$g_n(x) = \sum_{n=1}^{\infty} E_{mn} \sin \frac{m\pi x}{a}$$

where

$$E_{mn} = \frac{2}{a} \int_0^a g_n(x) \sin \frac{m\pi x}{a} \, dx.$$

If we also introduce the notation

$$\phi_m = \begin{cases} 0 \text{ when m = even} \\ 1 \text{ when m = odd} \end{cases}$$

then Eq. (3.64) can be expressed in the form

$$\sum_{m=1}^{\infty} \left[\phi_m \frac{dR_m}{dy} \bigg|_{y=b} + M_m \frac{df_m}{dy} \bigg|_{y=b} \right.$$

$$\left. + \sum_{n=1}^{\infty} E_{mn} (-1)^n \frac{n\pi}{b} V_n \right] \sin \frac{m\pi x}{a} = 0 \qquad (3.65)$$

where we now recognize the common factor to be sin $m\pi x/a$.
Since Eq. (3.65) must be satisfied independent of the position x

$$\phi_m \left(\frac{dR_m}{dy}\right)_{y=b} + M_m \left(\frac{df_m}{dy}\right)_{y=b}$$

$$+ \sum_{n=1}^{\infty} E_{mn} (-1)^n \left(\frac{n\pi}{b}\right) V_n = 0. \qquad (3.66)$$

In this problem it is not possible to obtain a general expression for M_m or V_n. Instead, Eq. (3.66) yields an infinite set of equations in the unknowns M_m and V_n. As an example, suppose that an approximate value of the deflection is to be obtained by a truncation of the series after m = n = 2. Then Eq. (3.66) becomes

$$\left(\frac{dR_1}{dy}\right)_{y=b} + M_1 \left(\frac{df_1}{dy}\right)_{y=b} - \frac{\pi}{b} E_{11} V_1 + \frac{2\pi}{b} E_{12} V_2 = 0$$

$$M_2 \left(\frac{df_2}{dy}\right)_{y=b} - \frac{\pi}{b} E_{21} V_1 + \frac{2\pi}{b} E_{22} V_2 = 0.$$

In a similar manner, two additional equations are obtained from the zero shear boundary condition given by Eq. (3.63b). The simultaneous solution of the resulting four equations yields the four unknowns M_1, M_2, V_1, and V_2. For a more accurate solution we would truncate after a larger value of m or n or both.

If convergence is slow, we must use many terms to obtain an accurate representation of the deflection. In that case, the set of simultaneous algebraic equations which must be solved for the constants, M_m and V_n, becomes large. Numerous techniques are available with which we can obtain solutions to large sets of linear algebraic equations.

Many important problems have been solved using the method of superposition. As we have seen, the manipulations required to obtain the solution are somewhat lengthy and the final solution appears to be cumbersome. However, once we obtain the general solution the digital computer can again be of considerable assistance.

The reader should not be misled by the difficulties involved in this technique into thinking that such solutions are too involved to be feasible. If we think of these solutions in the context of working several problems for a single homework assignment, then certainly it becomes a large task. However, when we encounter such a problem on the job (i.e., in a full-time work environment), the time and effort required for the solution is generally not unreasonable.

In this chapter we have examined the classic techniques which lead to the lateral deflection of rectangular plates. Many such problems have been solved previously and the results are tabulated in various reference books.[1,14]

Problems

14. Carefully explain why Eq. (3.5) can be reduced to Eq. (3.6).

15. Consider a simply supported rectangular steel plate under the action of the transverse load p = 20x psi. Use the Navier technique to determine the lateral deflection, w, at the midpoint. Assume the following dimensions.

$$a = 10 \text{ in}$$
$$b = 20 \text{ in}$$
$$h = .25 \text{ in}$$

Set up expressions for w

Set up solution (see eqtn 3-16)

16. Determine the lateral deflection of the midpoint of the steel plate in Fig. 3.8. Assume the plate thickness is .25 in, the loads are in pounds, and all edges are simply supported.

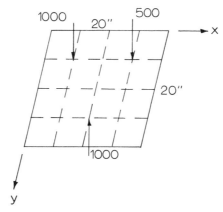

Fig. 3.8

17. Use the concept of influence coefficients to determine an expression for the deflections of the simply supported thin plate in Fig. 3.9. The units of a and b are inches.

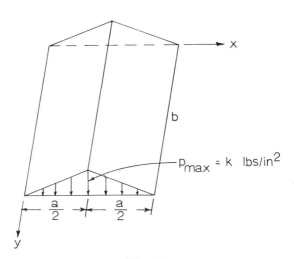

Fig. 3.9

18. Derive an expression for the lateral deflections of the thin plates in Fig. 3.10.

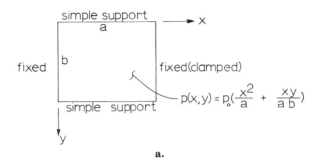

a.

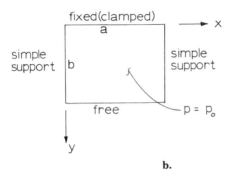

b.

Fig. 3.10

19. Derive an expression for the lateral deflections of the uniformly loaded plate in Fig. 3.11.

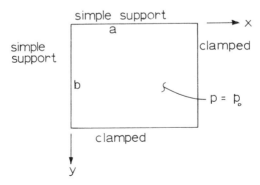

Fig. 3.11

Chapter 4

CLASSICAL SOLUTIONS FOR THE LATERAL DEFLECTIONS OF CIRCULAR PLATES

4.1 CIRCULAR PLATES WITH ARBITRARY LOADS

It is important that solutions be developed which lead to the lateral deflections of circular plates, since circular plates occur frequently in many structures. The general procedure for the determination of the lateral deflections of circular plates follows the same pattern that we employed with rectangular plates in Chapter 3. The expression for the lateral deflections of a circular plate is taken as the sum of the homogeneous part and the particular part of the governing equation

$$\nabla^4 w = \frac{p(r,\theta)}{D} \tag{4.1}$$

where the lateral load p is given in terms of the polar coordinates r and θ.

The polar coordinate system used for circular plates is shown in Fig. 4.1.

We begin the derivation of the solution of Eq. (4.1) by assuming the homogeneous part as

$$w_h = \sum_{n=0}^{\infty} f_n(r) \cos n\theta + \sum_{n=1}^{\infty} g_n(r) \sin n\theta \tag{4.2}$$

83

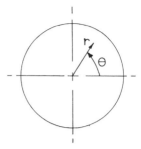

Fig. 4.1.   Polar coordinate system for a circular plate

where the functions $f_n(r)$ and $g_n(r)$ are determined as a result of satisfying the governing equation and the boundary conditions. The trigonometric functions $\cos n\theta$ and $\sin n\theta$ enforce the requirement that the expression for the lateral deflection must be single valued. Since these trigonometric functions are periodic in the interval $\theta = 2\pi$, that is

$$w(r,\theta) = w(r,\theta + 2n\pi) \qquad \text{for } n = 1, 2, 3, \ldots$$

the condition of single valuedness is met.

Substitution of Eq. (4.2) into the homogeneous equation

$$\nabla^4 w_h = 0$$

yields

$$
\nabla^4 w_h = \sum_{n=0}^{\infty} \left[\frac{d^4 f_n}{dr^4} + \frac{2}{r} \frac{d^3 f_n}{dr^3} - \frac{1 + 2n^2}{r^2} \frac{d^2 f_n}{dr^2} \right.
$$
$$
\left. + \frac{1 + 2n^2}{r^3} \frac{df_n}{dr} + \frac{n^2(n-4)}{r^4} f_n \right] \cos n\theta
$$
$$
+ \sum_{n=1}^{\infty} \left[\frac{d^4 g_n}{dr^4} + \frac{2}{r} \frac{d^3 g_n}{dr^3} - \frac{1 + 2n^2}{r^2} \frac{d^2 g_n}{dr^2} \right.
$$
$$
\left. + \frac{1 + 2n^2}{r^3} \frac{dg_n}{dr} + \frac{n^2(n^2-4)}{r^4} g_n \right] \sin n\theta = 0. \tag{4.3}
$$

Since Eq. (4.3) must equal zero for all values of r and θ

$$
\frac{d^4 f_n}{dr^4} + \frac{2}{r} \frac{d^3 f_n}{dr^3} - \frac{1 + 2n^2}{r^2} \frac{d^2 f_n}{dr^2} + \frac{1 + 2n^2}{r^3} \frac{df_n}{dr}
$$
$$
+ \frac{n^2(n^2-4)}{r^4} f_n = 0, \qquad n = 0, 1, 2, \ldots, \infty \tag{4.4}
$$

and

$$\frac{d^4 g_n}{dr^4} + \frac{2}{r} \frac{d^3 g_n}{dr^3} - \frac{1 + 2n^2}{r^2} \frac{d^2 g_n}{dr^2} + \frac{1 + 2n^2}{r^3} \frac{dg_n}{dr}$$
$$+ \frac{n^2(n^2 - 4)}{r^4} g_n = 0, \qquad n = 1, 2, 3, \ldots, \infty \qquad (4.5)$$

Equations (4.4) and (4.5) are equidimensional ordinary differential equations. They can be solved by solutions of the type[13]

$$f_n(r) = b_n r^\lambda, \qquad n = 0, 1, 2, \ldots, \infty \qquad (4.6a)$$

and

$$g_n(r) = c_n r^\lambda, \qquad n = 1, 2, 3, \ldots, \infty \qquad (4.6b)$$

in which b_n and c_n are arbitrary constants. If we substitute Eqs. (4.6a) and (4.6b) into Eqs. (4.4) and (4.5) respectively, we obtain the following characteristic equation in each case.

$$\lambda(\lambda - 1)(\lambda - 2)(\lambda - 3) + 2\lambda(\lambda - 1)(\lambda - 2)$$
$$- (1 + 2n^2)\lambda(\lambda - 1) + (1 + 2n^2)\lambda + n^2(n^2 - 4) = 0 \qquad (4.7)$$

The roots of Eq. (4.7) are

$$\lambda_1 = n, \quad \lambda_2 = -n, \quad \lambda_3 = n + 2, \quad \lambda_4 = -n + 2.$$

The total solution to Eqs. (4.4) and (4.5) must include all possible roots of Eq. (4.7). When $n = 0$, the roots λ_1, λ_2, λ_3, and λ_4 correspond to the function $f_n(r)$ only. Setting $n = 0$ leads us to the following two sets of double roots.

$$\lambda_1 = \lambda_2 = 0, \quad \lambda_3 = \lambda_4 = 2$$

From the theory of equidimensional equations, we know that a double root leads to the terms r^λ and $(\ln r)r^\lambda$, where ln is logarithm to the base e. Thus, we have

$$f_0(r) = A_0 r^0 + B_0 r^2 + C_0 r^0 \ln r + D_0 r^2 \ln r$$
$$= A_0 + B_0 r^2 + C_0 \ln r + D_0 r^2 \ln r. \qquad (4.8)$$

The functions $f_n(r)$ and $g_n(r)$ both exist when $n = 1$. The corresponding roots of Eq. (4.7) are

$$\lambda_1 = \lambda_4 = 1$$

$$\lambda_2 = -1$$

$$\lambda_3 = 3.$$

In this case, one set of double roots exists and we have

$$f_1 = A_1 r + B_1 r^3 + C_1 r^{-1} + D_1 r \ln r \qquad (4.9a)$$

and

$$g_1 = A_1' r + B_1' r^3 + C_1' r^{-1} + D_1' r \ln r. \qquad (4.9b)$$

For all values of n greater than one, the functions $f_n(r)$ and $g_n(r)$ exist with four distinct roots. Thus, for $n > 1$ these functions take the form

$$f_n = A_n r^n + B_n r^{-n} + C_n r^{n+2} + D_n r^{-n+2} \qquad (4.10a)$$

and

$$g_n = A_n' r^n + B_n' r^{-n} + C_n' r^{n+2} + D_n' r^{-n+2}. \qquad (4.10b)$$

Equation (4.3) becomes

$$\begin{aligned}
w_h = {}& A_0 + B_0 r^2 + C_0 \ln r + D_0 r^2 \ln r \\
& + (A_1 r + B_1 r^3 + C_1 r^{-1} + D_1 r \ln r) \cos \theta \\
& + (A_1' r + B_1' r^3 + C_1' r^{-1} + D_1' r \ln r) \sin \theta \\
& + \sum_{n=2}^{\infty} (A_n r^n + B_n r^{-n} + C_n r^{n+2} + D_n r^{-n+2}) \cos n\theta \\
& + \sum_{n=2}^{\infty} (A_n' r^n + B_n' r^{-n} + C_n' r^{n+2} + D_n' r^{-n+2}) \sin n\theta. \qquad (4.11)
\end{aligned}$$

Next we shall develop the particular solution. The general technique for determination of the particular solution follows the same basic pattern that was prescribed in Chapter 3 for the Levy solution of rectangular plates. We begin by defining the particular solution as

$$w_p = F_0(r) + \sum_{n=1}^{\infty} [F_n(r) \cos n\theta + G_n(r) \sin n\theta]. \qquad (4.12)$$

Then the applied load is expanded in a Fourier series as follows.

$$p(r,\theta) = p_0(r) + \sum_{n=1}^{\infty} [P_n(r) \cos n\theta + S_n(r) \sin n\theta] \qquad (4.13)$$

in which

$$P_n(r) = \frac{1}{\pi} \int_{-\pi}^{\pi} p(r,\theta) \cos n\theta \, d\theta, \qquad n = 0, 1, 2, \ldots, \infty$$

$$S_n(r) = \frac{1}{\pi} \int_{-\pi}^{\pi} p(r,\theta) \sin n\theta \, d\theta, \qquad n = 1, 2, 3, \ldots, \infty.$$

Substitution of Eqs. (4.12) and (4.13) into Eq. (4.1) gives

$$\frac{d^4 F_0}{dr^4} + \frac{2}{r} \frac{d^3 F_0}{dr^3} - \frac{1}{r^2} \frac{d^2 F_0}{dr^2} + \frac{1}{r^3} \frac{dF_0}{dr}$$

$$+ \sum_{n=1}^{\infty} \left[\frac{d^4 F_n}{dr^4} + \frac{2}{r} \frac{d^3 F_n}{dr^3} - \frac{1+2n^2}{r^2} \frac{d^2 F_n}{dr^2} \right.$$

$$+ \left. \frac{1+2n^2}{r^3} \frac{dF_n}{dr} + \frac{n^2(n^2-4)}{r^4} F_n \right] \cos n\theta$$

$$+ \sum_{n=1}^{\infty} \left[\frac{d^4 G_n}{dr^4} + \frac{2}{r} \frac{d^3 G_n}{dr^3} - \frac{1+2n^2}{r^2} \frac{d^2 G_n}{dr^2} \right.$$

$$+ \left. \frac{1+2n^2}{r^3} \frac{dG_n}{dr} + \frac{n^2(n^2-4)}{r^4} G_n \right] \sin n\theta$$

$$= \frac{p_0(r)}{D} + \frac{1}{D} \sum_{n=1}^{\infty} p_n \cos n\theta + \frac{1}{D} \sum_{n=1}^{\infty} S_n \sin n\theta. \qquad (4.14)$$

Equation (4.14) is satisfied for all values of r and θ if

$$\frac{d^4 F_0}{dr^4} + \frac{2}{r} \frac{d^3 F_0}{dr^3} - \frac{1}{r^2} \frac{d^2 F_0}{dr^2} + \frac{1}{r^3} \frac{dF_0}{dr} = \frac{p_0(r)}{D} \qquad (4.15a)$$

$$\frac{d^4 F_n}{dr^4} + \frac{2}{r} \frac{d^3 F_n}{dr^3} - \frac{1+2n^2}{r^2} \frac{d^2 F_n}{dr^2} + \frac{1+2n^2}{r^2} \frac{dF_n}{dr}$$

$$+ \frac{n^2(n^2-4)}{r^4} F_n = \frac{P_n}{D} \qquad (4.15b)$$

and

$$\frac{d^4 G_n}{dr^4} + \frac{2}{r} \frac{d^3 G_n}{dr^3} - \frac{1+2n^2}{r^2} \frac{d^2 G_n}{dr^2} + \frac{1+2n^2}{r^2} \frac{dG_n}{dr}$$

$$+ \frac{n^2(n-4)}{r^4} G_n = \frac{S_n}{D}. \qquad (4.15c)$$

For a given lateral load, $p(r,\theta)$, the particular solution can be completed by solving the preceding three equidimensional differential equations for F_0, F_n, and G_n. The general expression for the total lateral deflection is the sum of Eqs. (4.11) and (4.12).

$$w = w_h + w_p$$

or

$$
\begin{aligned}
w = {} & A_0 + B_0 r^2 + C_0 \ln r + D_0 r^2 \ln r \\
& + (A_1 r + B_1 r^3 + C_1 r^{-1} + D_1 r \ln r) \cos \theta \\
& + (A_1' r + B_1' r^3 + C_1' r^{-1} + D_1' r \ln r) \sin \theta \\
& + \sum_{n=2}^{\infty} (A_n r^n + B_n r^{-n} + C_n r^{n+2} + D_n r^{-n+2}) \cos n\theta \\
& + \sum_{n=2}^{\infty} (A_n' r^n + B_n' r^{-n} + C_n' r^{n+2} + D_n' r^{-n+2}) \sin n\theta \\
& + F_0(r) + \sum_{n=1}^{\infty} [F_n(r) \cos n\theta + G_n(r) \sin n\theta].
\end{aligned}
\tag{4.16}
$$

The arbitrary constants (As, Bs, Cs, and Ds) are determined from prescribed boundary conditions. The procedure for determining these constants from a given set of boundary conditions is demonstrated if we consider the problem of a circular plate clamped at its inner and outer edges as shown in Fig. 4.2. The constants of Eq. (4.16) are determined from the boundary conditions

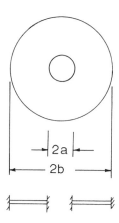

Fig. 4.2. Fully clamped plate

$$w \Big|_{r = a} = 0 \tag{4.17a}$$

$$w \Big|_{r = b} = 0 \tag{4.17b}$$

$$\frac{\partial w}{\partial r} \Big|_{r = a} = 0 \tag{4.17c}$$

$$\frac{\partial w}{\partial r} \Big|_{r = b} = 0. \tag{4.17d}$$

If we apply these boundary conditions to Eq. (4.16), we obtain the following four equations.

$$A_0 + B_0 a^2 + C_0 \ln a + D_0 a^2 \ln a$$
$$+ (A_1 a + B_1 a^3 + C_1 a^{-1} + D_1 a \ln a) \cos \theta$$
$$+ (A_1' a + B_1' a^3 + C_1' a^{-1} + D_1' a \ln a) \sin \theta$$

$$+ \sum_{n=2}^{\infty} (A_n a^n + B_n a^{-n} + C_n a^{n+2} + D_n a^{-n+2}) \cos n\theta$$

$$+ \sum_{n=2}^{\infty} (A_n' a^n + B_n' a^{-n} + C_n' a^{n+2} + D_n' a^{-n+2}) \sin n\theta + F_0(a)$$

$$+ \sum_{n=1}^{\infty} [F_n(a) \cos n\theta + G_n(a) \sin n\theta] = 0, \tag{4.18a}$$

$$A_0 + B_0 b^2 + C_0 \ln b + D_0 b^2 \ln b$$
$$+ (A_1 b + B_1 b^3 + C_1 b^{-1} + D_1 b \ln b) \cos \theta$$
$$+ (A_1' b + B_1' b^3 + C_1' b^{-1} + D_1' b \ln b) \sin \theta$$

$$+ \sum_{n=2}^{\infty} (A_n b^n + B_n b^{-n} + C_n b^{n+2} + D_n b^{-n+2}) \cos n\theta$$

$$+ \sum_{n=2}^{\infty} (A_n' b^n + B_n' b^{-n} + C_n' b^{n+2} + D_n' b^{-n+2}) \sin n\theta + F_0(b)$$

$$+ \sum_{n=1}^{\infty} [F_n(b) \cos n\theta + G_n(b) \sin n\theta] = 0, \tag{4.18b}$$

$$2B_0 a + C_0 a^{-1} + D_0(2a \ln a + a)$$
$$+ [A_1 + 3B_1 a^2 - C_1 a^{-2} + D_1(\ln a + 1)] \cos \theta$$
$$+ [A_1' + 3B_1' a^2 - C_1' a^{-2} + D_1'(\ln a + 1)] \sin \theta$$

$$+ \sum_{n=2}^{\infty} [nA_n a^{n-1} - nB_n a^{-n-1}$$

$$+ (n + 2) C_n a^{n+1} + (-n + 2) D_n a^{-n+1}] \cos n\theta$$

$$+ \sum_{n=2}^{\infty} [nA'_n a^{n-1} - nB'_n a^{-n-1}$$

$$+ (n + 2) C'_n a^{n+1} + (-n + 2) D'_n a^{-n+1}] \sin n\theta$$

$$+ \left\{ \frac{dF_0}{dr} + \sum_{n=1}^{\infty} \left[\frac{dF_n}{dr} \cos n\theta + \frac{dG_n}{dr} \sin n\theta \right] \right\}_{r = a} = 0, \quad (4.18c)$$

$$2B_0 b + C_0 b^{-1} + D_0(2b \ln b + b)$$
$$+ [A_1 + 3B_1 b^2 - C_1 b^{-2} + D_1(\ln b + 1)] \cos \theta$$
$$+ [A'_1 + 3B'_1 b^2 - C'_1 b^{-2} + D'_1(\ln b + 1)] \sin \theta$$

$$+ \sum_{n=2}^{\infty} [nA_n b^{n-1} - nB_n b^{-n-1}$$

$$+ (n + 2) C_n b^{n+1} + (-n + 2) D_n b^{-n+1}] \cos n\theta$$

$$+ \sum_{n=2}^{\infty} [nA'_n b^{n-1} - nB'_n b^{-n-1}$$

$$+ (n + 2) C'_n b^{n+1} + (-n + 2) D'_n b^{-n+1}] \sin n\theta$$

$$+ \left\{ \frac{dF_0}{dr} + \sum_{n=1}^{\infty} \left[\frac{dF_n}{dr} \cos n\theta + \frac{dG_n}{dr} \sin n\theta \right] \right\}_{r = b} = 0. \quad (4.18d)$$

If we carefully examine Eqs. (4.18), we see that each one is of the type

$$k_0 + k_1 \sin \theta + K_1 \cos \theta + \sum_{n=2}^{\infty} k_n \sin n\theta + \sum_{n=2}^{\infty} K_n \cos n\theta = 0 \quad (4.19)$$

where k_0, k_1, and K_1 are functions of either a or b, and k_n and K_n are functions of n and either a or b. For Eq. (4.19) to be satisfied for all values of θ, we must have

$$k_0 = 0 \qquad (4.20a)$$

$$k_1 = 0 \qquad (4.20b)$$

$$K_1 = 0 \qquad (4.20c)$$

$$k_n = 0 \qquad (4.20d)$$

$$K_n = 0. \qquad (4.20e)$$

If we apply this idea to the entire set of Eqs. (4.19), we obtain the following five sets of equations.

$$A_0 + B_0 a^2 + C_0 \ln a + D_0 a^2 \ln a + F_0(a) = 0 \qquad (4.21a)$$

$$A_0 + B_0 b^2 + C_0 \ln b + D_0 b^2 \ln b + F_0(b) = 0 \qquad (4.21b)$$

$$2B_0 a + C_0 a^{-1} + D_0 (a + 2a \ln a) + \left(\frac{dF_0}{dr}\right)_{r = a} = 0 \quad (4.21c)$$

$$2B_0 b + C_0 b^{-1} + D_0 (b + 2b \ln b) + \left(\frac{dF_0}{dr}\right)_{r = b} = 0 \quad (4.21d)$$

$$A_1' a + B_1' a^3 + C_1' a^{-1} + D_1' a \ln a + G_1(a) = 0 \qquad (4.22a)$$

$$A_1' b + B_1' b^3 + C_1' b^{-1} + D_1' b \ln b + G_1(b) = 0 \qquad (4.22b)$$

$$A_1' + 3B_1' a^2 - C_1' a^{-2} + D_1' (\ln a + 1) + \left(\frac{dG_1}{dr}\right)_{r = a} = 0 \qquad (4.22c)$$

$$A_1' + 3B_1' b^2 - C_1' b^{-2} + D_1' (\ln b + 1) + \left(\frac{dG_1}{dr}\right)_{r = b} = 0 \qquad (4.22d)$$

$$A_1 a + B_1 a^3 + C_1 a^{-1} + D_1 a \ln a + F_1(a) = 0 \qquad (4.23a)$$

$$A_1 b + B_1 b^3 + C_1 b^{-1} + D_1 b \ln b + F_1(b) = 0 \qquad (4.23b)$$

$$A_1 + 3B_1 a^2 - C_1 a^{-2} + D_1 (1 + \ln a) + \left(\frac{dF_1}{dr}\right)_{r = a} = 0 \qquad (4.23c)$$

$$A_1 + 3B_1 b^2 - C_1 b^{-2} + D_1 (1 + \ln b) + \left(\frac{dF_1}{dr}\right)_{r = b} = 0 \qquad (4.23d)$$

$$A_n' a^n + B_n' a^{-n} + C_n' a^{n+2} + D_n' a^{-n+2} + G_n(a) = 0 \qquad (4.24a)$$

$$A_n' b^n + B_n' b^{-n} + C_n' b^{n+2} + D_n' b^{-n+2} + G_n(b) = 0 \qquad (4.24b)$$

$$nA_n' a^{n-1} - nB_n' a^{-n-1} + (n + 2) C_n' a^{n+1} + (-n + 2) D_n' a^{-n+1}$$
$$+ \left(\frac{dG_n}{dr}\right)_{r = a} = 0 \qquad (4.24c)$$

$$nA_n' b^{n-1} - nB_n' b^{-n-1} + (n + 2) C_n' b^{n+1} + (-n + 2) D_n' b^{-n+1}$$
$$+ \left(\frac{dG_n}{dr}\right)_{r = b} = 0 \qquad (4.24d)$$

$$A_n a^n + B_n a^{-n} + C_n a^{n+2} + D_n a^{-n+2} + F_n(a) = 0 \qquad (4.25a)$$

$$A_n b^n + B_n b^{-n} + C_n b^{n+2} + D_n b^{-n+2} + F_n(b) = 0 \qquad (4.25b)$$

$$nA_n a^{n-1} - nB_n a^{-n-1} + (n + 2) C_n a^{n+1} + (-n + 2) D_n a^{-n+1}$$
$$+ \left(\frac{dF_n}{dr}\right)_{r = a} = 0 \qquad (4.25c)$$

$$nA_n b^{n-1} - nB_n b^{-n-1} + (n + 2) C_n b^{n+1} + (-n + 2) D_n b^{-n+1}$$
$$+ \left(\frac{dF_n}{dr}\right)_{r = b} = 0 \qquad (4.25d)$$

Equations (4.21) are analogous to Eq. (4.20a); they represent the constant terms. Equations (4.22) are analogous to Eq. (4.20b); they represent the terms that are coefficients of sin θ. Equations (4.23) are analogous to Eq. (4.20c); they represent the terms that are coefficients of cos θ. Similarly, Eqs. (4.24) and (4.25), which are analogous to Eqs. (4.20d) and (4.20e) respectively, are the coefficients of sin $n\theta$ and cos $n\theta$ respectively.

Equations (4.21), (4.22), and (4.23) are sets of linear algebraic equations, each of which can be solved simultaneously for their respective set of constants. Equations (4.24) can be solved simultaneously for A_2', B_2', C_2', and D_2' if n = 2. The same equations can next be solved for A_3', B_3', C_3', and D_3' by setting n = 3. We repeat this procedure for Eqs. (4.24) and (4.25) until the lateral deflections of Eq. (4.16) converge to a satisfactory solution.

Although this procedure appears cumbersome, it is readily adapted to the digital computer.

4.2   CIRCULAR PLATES OF SYMMETRY

If the boundary conditions and external loads for a circular plate are symmetric about the z axis, Fig. 2.1, the lateral deflections are also symmetric about that axis. The expression defining the lateral deflections for such a plate cannot be a function of the independent variable θ. Consequently, the total lateral deflection given by Eq. (4.16) is reduced to

$$w = A_0 + B_0 r^2 + C_0 \ln r + D_0 r^2 \ln r + F_0(r). \qquad (4.26)$$

The first four terms of Eq. (4.26) represent the homogeneous solution of the governing equation and may be obtained by eliminating from Eq. (4.11) those terms that are functions of θ. The last term of Eq. (4.26) represents the particular solution. It is a consequence of eliminating from Eq. (4.12) the terms dependent upon θ, and may be determined from Eq. (4.15a).

As an example, let's consider a simply supported circular plate under the action of a uniform load p_0 as shown in Fig. 4.3.

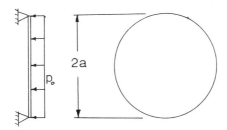

Fig. 4.3. Simply supported circular plate with uniform load p_0

The particular solution may be determined from Eq. (4.15a), where $p_0(r)$ becomes p_0.

$$\frac{d^4 F_0}{dr^4} + \frac{2}{r}\frac{d^3 F_0}{dr^3} - \frac{1}{r^2}\frac{d^2 F_0}{dr^2} + \frac{1}{r^3}\frac{dF_0}{dr} = \frac{p_0}{D} \tag{4.27}$$

or

$$r^4 \frac{d^4 F_0}{dr^4} + 2r^3 \frac{d^3 F_0}{dr^3} - r^2 \frac{d^2 F_0}{dr^2} + r \frac{dF_0}{dr} = \frac{p_0}{D} r^4 \tag{4.28}$$

Equation (4.28) is an equidimensional equation whose particular solution is readily determined as

$$F_0 = \frac{p_0 r^4}{64D}. \tag{4.29}$$

The total solution now becomes

$$w = A_0 + B_0 r^2 + C_0 \ln r + D_0 r^2 \ln r + \frac{p_0 r^4}{64D}. \tag{4.30}$$

The lateral deflections described by Eq. (4.30) must remain finite at

$r = 0$; hence, we must require

$$C_0 = D_0 = 0$$

which reduces Eq. (4.30) to

$$w = A_0 + B_0 r^2 + \frac{p_0 r^4}{64D}. \tag{4.31}$$

The boundary conditions for a simply supported plate are given in Eqs. (2.15a) and (2.15b) as

$$w \Big|_{r = a} = 0 \tag{4.32a}$$

and

$$\left(\frac{d^2w}{dr^2} + \frac{\nu}{r} \frac{dw}{dr} \right)_{r = a} = 0. \tag{4.32b}$$

If we apply these boundary conditions to Eq. (4.31), we find

$$A_0 + B_0 a^2 + \frac{p_0 a^4}{64D} = 0 \tag{4.33a}$$

and

$$2B_0(1 + \nu) + \frac{a^2}{16D} (3 + \nu) = 0. \tag{4.33b}$$

The simultaneous solution of Eqs. (4.33a) and (4.33b) results in

$$A_0 = \frac{a^4}{32D} \left(\frac{3 + \nu}{1 + \nu} - \frac{p_0}{2D} \right)$$

and

$$B_0 = - \frac{a^2}{32D} \left(\frac{3 + \nu}{1 + \nu} \right).$$

Thus, the final expression defining the lateral deflections of the simply supported plate in Fig. 4.3 is

$$w = \frac{a^4}{32D} \left(\frac{3 + \nu}{1 + \nu} - \frac{p_0}{2D} \right) - \frac{a^2}{32D} \left(\frac{3 + \nu}{1 + \nu} \right) r^2 + \frac{p_0 r^4}{64D}. \tag{4.34}$$

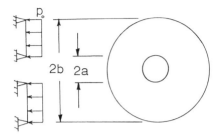

Fig. 4.4. Simply supported circular plate with uniform load p_0

As an additional example, suppose that the plate now has a hole which results in an inner boundary that is also simply supported, as shown in Fig. 4.4.

The particular solution may again be determined from Eq. (4.15a) as

$$F_0 = \frac{p_0 r^4}{64D} \tag{4.35}$$

and the total solution once again becomes

$$w = A_0 + B_0 r^2 + C_0 \ln r + D_0 r^2 \ln r + \frac{p_0 r^4}{64D}. \tag{4.36}$$

The boundary conditions according to Eqs. (2.15) are

$$w \Big|_{r = a} = 0 \tag{4.37a}$$

$$w \Big|_{r = b} = 0 \tag{4.37b}$$

$$\left(\frac{\partial^2 w}{\partial r^2} + \frac{\nu}{r} \frac{\partial w}{\partial r} \right)_{r = a} = 0 \tag{4.37c}$$

$$\left(\frac{\partial^2 w}{\partial r^2} + \frac{\nu}{r} \frac{\partial w}{\partial r} \right)_{r = b} = 0. \tag{4.37d}$$

If we apply these boundary conditions to Eq. (4.36), we obtain

$$A_0 + B_0 a^2 + C_0 \ln a + D_0 a^2 \ln a + \frac{p_0 a^4}{64D} = 0 \tag{4.38a}$$

$$A_0 + B_0 b^2 + C_0 \ln b + D_0 b^2 \ln b + \frac{p_0 b^4}{64D} = 0 \qquad (4.38b)$$

$$2B_0(1 + \nu) - \frac{C_0}{a^2} (1 - \nu) + D_0[3 + 2\ln a + \nu(2\ln a + 1)]$$

$$+ \frac{p_0 a^2}{16D} (3 + \nu) = 0 \qquad (4.38c)$$

$$2B_0(1 + \nu) - \frac{C_0}{b^2} (1 - \nu) + D_0[3 + 2\ln b + \nu(2\ln b + 1)]$$

$$+ \frac{p_0 b^2}{16D} (3 + \nu) = 0. \qquad (4.38d)$$

The simultaneous solution of Eqs. (4.38) yields the constants A_0, B_0, C_0, and D_0, and thus completes the solution.

Problems

20. Determine an expression for the lateral deflections of the plate in Fig. 4.3 if the edge is clamped.

21. Determine the lateral deflection of the midpoint of the simply supported plate in Fig. 4.5.

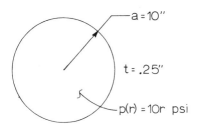

Fig. 4.5

22. Derive an expression for the lateral deflections of the thin plate in Fig. 4.2. Assume the load is a uniform 10 psi, $a = 2$ in, $b = 8$ in, and the material is steel. Calculate the required thickness to maintain $w_{max} \le .06$ in.

Chapter 5
CONTINUOUS PLATES AND PLATES ON ELASTIC FOUNDATIONS

5.1 INTRODUCTION

Structures often consist of a single plate supported by intermediate beams or columns as illustrated in Figs. 5.1 and 5.2. A plate of this type is referred to as a continuous plate.

In this chapter, we will show that continuous plates can be analyzed by subdividing the plate into smaller component plate sections bounded by the intermediate supporting beams or columns. Figure 5.3 illustrates the manner in which the plates in Figs. 5.1 and 5.2 would be subdivided.

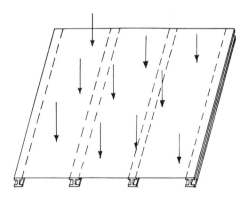

Fig. 5.1. Rectangular plate supported by intermediate beams

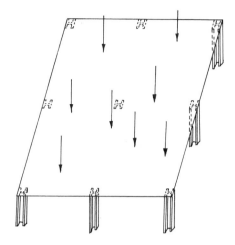

Fig. 5.2.   Rectangular plate supported by intermediate columns

These component plates are first analyzed individually. Then the solutions for lateral deflections of the component plates are forced to satisfy the appropriate conditions of continuity along the common internal boundaries. In addition to this method of solution, continuous plates can be analyzed using the methods presented in Chapters 7, 8, and 9.

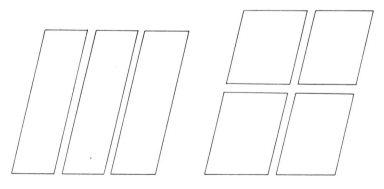

a.   Component plates in Fig. 5.1   b.   Component plates in Fig. 5.2

Fig. 5.3.   Subdivision of continuous plates into component plates

5.2 PLATES WITH INTERMEDIATE BEAMS WHICH PROHIBIT DEFLECTIONS

As a first example of continuous plates, consider the plate in Fig. 5.4. . In this example we will assume that the supporting I beams are very rigid in comparison with the plate. Thus, we can assume that the plate has zero deflection along the centerline of the supporting beams. Let us also assume that the plate and beams are connected such that the beams do not restrict rotation of the plate. A single row of fasteners along the beams with slightly oversized holes might lead to such a situation, since for small deflections only a very small amount of rotation at the fasteners is necessary for the plate to behave as if it is simply supported. We will assume that each supporting beam has no width, and

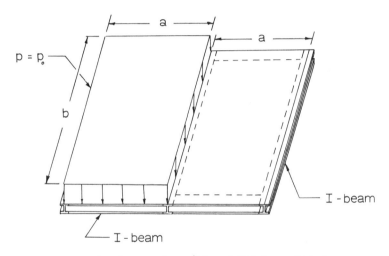

Fig. 5.4. Continuous plate with a rigid intermediate beam

consequently exists only along its centerline. This assumption is valid if the cross-sectional dimensions of the beam are small compared to the plate dimensions a and b.

To solve this problem, we examine the component plates in Fig. 5.5

The simply supported boundary conditions on three sides of each of the component plates precisely match the conditions of the total continuous plate. The boundary condition of deflection, w, equal to zero at $x_1 = a$ and $x_2 = 0$ also matches the condition of the total continuous plate. The choice of the second boundary condition along the common edge requires considerably more insight into the problem. We observe that this boundary condition has been specified as the bending moment,

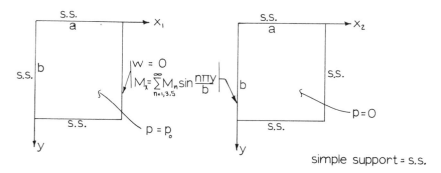

a. Component plate 1   b. Component plate 2

Fig. 5.5.   Subdivided continuous plate

and that we have represented this bending moment function by a Fourier sine series with an unknown set of coefficients M_n. We can better understand this boundary condition by realizing that our procedure will be first to determine separate solutions for plates 1 and 2 each as a function of M_n, and then to determine the necessary bending moment function, i.e. the values of the coefficients M_n which will make the two solutions compatible along this common edge. We should note that assigning the same moment function to the common face of both plates satisfies the assumption that rotation is not resisted by the intermediate beam. In other words, we say that the variation of bending moment must be continuous; and since the support does not resist plate bending, the bending moment must be the same on both sides of the supporting beam.

At this point, we might wonder why the Fourier series is selected. We recall that a piecewise continuous function existing from $0 \leq y \leq b$ can be represented by a Fourier series. Consequently, we know that the exact representation of M_x is available to us if we are able to determine the proper infinite set of M_n's. Since the lateral load and the boundary conditions are symmetric about a line $y = b/2$, the moment function along the edges $x_1 = a$ and $x_2 = 0$ must also be symmetric about $y = b/2$. Thus, we discard all even values of n, since they lead to antisymmetric functions. The reader may now wonder why a sine series has been selected rather than a cosine series or some other kind of series. In the process of solving this problem, the reason for this choice will be disclosed at the appropriate time.

To continue the analysis, we observe that the lateral deflections of each of the two component plates in Fig. 5.5 can be determined by the Levy technique. If we refer to Eq. (3.43), and replace y by x_1, x by y, and a by b, we can write the general solution for Plate 1 in terms of the x_1 and y coordinates as follows.

$$w_1 = \sum_{n=1,3,5,\ldots}^{\infty} \left(A_n \sinh \frac{n\pi x_1}{b} + B_n \cosh \frac{n\pi x_1}{b} + C_n x_1 \sinh \frac{n\pi x_1}{b} \right.$$

$$\left. + D_n x_1 \cosh \frac{n\pi x_1}{b} \right) \sin \frac{n\pi y}{b} + \sum_{n=1,3,5,\ldots}^{\infty} \frac{4p_0 b^4}{n^5 \pi^5 D} \sin \frac{n\pi y}{b} \tag{5.1}$$

The appropriate boundary conditions are

$$w_1 \Big|_{x_1 = 0} = 0 \tag{5.2a}$$

$$\frac{\partial^2 w_1}{\partial x_1^2} \Bigg|_{x_1 = 0} = 0 \tag{5.2b}$$

$$w_1 \Big|_{x_1 = a} = 0 \tag{5.2c}$$

$$-D \frac{\partial^2 w_1}{\partial x_1^2} \Bigg|_{x_1 = a} = \sum_{n=1,3,5,\ldots}^{\infty} M_n \sin \frac{n\pi y}{b} . \tag{5.2d}$$

Substitution of Eq. (5.1) into the boundary conditions specified by Eqs. (5.2a), (5.2b), (5.2c), and (5.2d) yields the following four linear algebraic equations in the four unknown coefficients A_n, B_n, C_n, and D_n.

$$B_n + \frac{4p_0 b^4}{n^5 \pi^5 D} = 0 \tag{5.3a}$$

$$\frac{n\pi}{b} B_n + 2C_n = 0 \tag{5.3b}$$

$$A_n \sinh \frac{n\pi a}{b} + B_n \cosh \frac{n\pi a}{b} + C_n a \sinh \frac{n\pi a}{b}$$

$$+ D_n a \cosh \frac{n\pi a}{b} + \frac{4p_0 b^4}{n^5 \pi^5 D} = 0 \tag{5.3c}$$

$$A_n \sinh \frac{n\pi a}{b} + B_n \cosh \frac{n\pi a}{b} + C_n \left[\frac{2b}{n\pi} \cosh \frac{n\pi a}{b} + a \sinh \frac{n\pi a}{b} \right]$$

$$+ D_n \left[\frac{2b}{n\pi} \sinh \frac{n\pi a}{b} + a \cosh \frac{n\pi a}{b} \right] + \frac{M_n b^2}{D n^2 \pi^2} = 0 \tag{5.3d}$$

At this point, we may direct attention to the reason for representing the moment along the common edge by a Fourier sine series. When Eq. (5.1) is substituted into Eq. (5.2d), all terms are infinite summations and are multiplied by sin $(n\pi y/b)$. If this equation is to be satisfied for all values of y, we are able to reduce it to the algebraic equation given by Eq. (5.3d). If the bending moment is represented by a function other than a Fourier sine series, mathematical complications that we like to avoid would occur. From the simultaneous solution of Eqs. (5.3a), (5.3b), (5.3c), and (5.3d) we obtain

$$B_n = -\frac{4p_0 b^4}{n^5 \pi^5 D} \qquad (5.4a)$$

$$C_n = \frac{2p_0 b^3}{n^4 \pi^4 D} \qquad (5.4b)$$

$$D_n = -\frac{\operatorname{csch} \frac{n\pi a}{b}}{D} \left[M_n \frac{b}{2n\pi} + \frac{2p_0 b^3}{n^4 \pi^4} \left(-1 + \cosh \frac{n\pi a}{b} \right) \right] \qquad (5.4c)$$

$$A_n = \frac{1}{D} \left\{ a \frac{\coth \frac{n\pi a}{b}}{\sinh \frac{n\pi a}{b}} \left[M_n \frac{b}{2n\pi} + \frac{2p_0 b^3}{n^4 \pi^4} \left(-1 + \cosh \frac{n\pi a}{b} \right) \right] \right.$$
$$\left. + \frac{4p_0 b^4}{n^5 \pi^5} \left(\coth \frac{n\pi a}{b} - \frac{n\pi a}{2b} - \operatorname{csch} \frac{n\pi a}{b} \right) \right\}. \qquad (5.4d)$$

We now have w in terms of the applied load, the plate geometry, the material properties, and the unknown coefficient M_n. Before we can determine M_n we must derive the corresponding expression for w_2 for the component Plate 2. If we examine Fig. 5.5, it becomes apparent that the expression for w_2 is very similar to the expression for w_1 given by Eq. (5.1). The only difference in the general form is that w_2 does not contain a particular solution because of the absence of a lateral load. Thus, the expression for w_2, written in terms of the coordinates x_2 and y and, of course, a different set of coefficients, is

$$w_2 = \sum_{n=1,3,5,\ldots}^{\infty} \left[K_n \sinh \frac{n\pi x_2}{b} + L_n \cosh \frac{n\pi x_2}{b} \right.$$
$$\left. + N_n x_2 \sinh \frac{n\pi x_2}{b} + Q_n x_2 \cosh \frac{n\pi x_2}{b} \right] \sin \frac{n\pi y}{b}. \qquad (5.5)$$

The boundary conditions which must be satisfied by w_2 are

$$w_2 \bigg|_{x_2 = 0} = 0$$

$$\frac{\partial^2 w_2}{\partial x_2^2} \bigg|_{x_2 = a} = 0$$

$$w_2 \bigg|_{x_2 = a} = 0$$

$$-D \frac{\partial^2 w_2}{\partial x_2^2} \bigg|_{x_2 = 0} = \sum_{n=1,3,5,\ldots}^{\infty} M_n \sin \frac{n\pi y}{b}$$

and lead to the following expressions.

$$L_n = 0 \tag{5.6a}$$

$$K_n = -\frac{M_n ab}{2Dn\pi} \left[1 + \left(\coth \frac{n\pi a}{b} \right)^2 \right] \tag{5.6b}$$

$$Q_n = -\frac{M_n b}{2Dn\pi} \coth \frac{n\pi a}{b} \tag{5.6c}$$

$$N_n = \frac{M_n b}{2Dn\pi} \tag{5.6d}$$

Now we have w_2 as a function of the plate geometry, the material constants, and the unknown coefficient M_n. We determine M_n from the following condition of continuity.

$$\frac{\partial w_1}{\partial x_1} \bigg|_{x_1 = a} = \frac{\partial w_2}{\partial x_2} \bigg|_{x_2 = 0} \tag{5.7}$$

Equation (5.7) leads directly to the expression

$$A_n \frac{n\pi}{b} \cosh \frac{n\pi a}{b} + B_n \frac{n\pi}{b} \sinh \frac{n\pi a}{b}$$
$$+ C_n \left(\sinh \frac{n\pi a}{b} + \frac{n\pi a}{b} \cosh \frac{n\pi a}{b} \right)$$
$$+ D_n \left(\cosh \frac{n\pi a}{b} + \frac{n\pi a}{b} \sinh \frac{n\pi a}{b} \right) - K_n \frac{n\pi}{b} - Q_n = 0. \tag{5.8}$$

Substitution of the values for A_n, B_n, C_n, D_n, K_n, and Q_n from Eqs. (5.4) and (5.6) into Eq. (5.8) leads to the solution for M_n, and the solution to our problem is now complete. We again find that the form of our solution will contain many terms; however, this type of solution is particularly adaptable to the digital computer.

5.3 PLATES WITH INTERMEDIATE ELASTIC SUPPORTS WHICH RESIST LATERAL DEFLECTIONS

A second type of continuous plate is encountered when the center support in Fig. 5.4 is relatively flexible compared with the flexibility of the plate. In this case, the deflection of the plate is not zero along the center support, but is a function of the stiffness of the support. As before, it is assumed that the intermediate support does not resist rotation of the plate.

To solve this problem, we again subdivide the continuous plate into two plates as shown in Fig. 5.6. The only change from the plate in Section 5.2 is in the boundary conditions along the common edge.

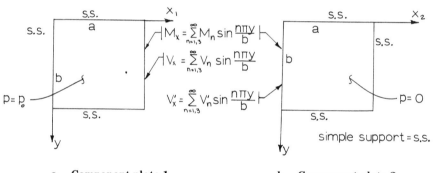

a. Component plate 1   b. Component plate 2

Fig. 5.6. Subdivided continuous plate with an intermediate elastic support

We choose Fourier sine series to represent M_x and V_x for the reasons presented in Section 5.2. Again M_x is defined by the same function on both sides of the common edge, since the support does not resist rotation of the plate. However, we have allowed the shear, V_x, to be different on opposite sides of the elastic support because the support carries some of the shear load.

The general solutions for lateral deflections are again described by Eqs. (5.1) and (5.5) and are rewritten here.

$$w_1 = \sum_{n=1,3,5,...}^{\infty} \left(A_n \sinh \frac{n\pi x_1}{b} + B_n \cosh \frac{n\pi x_1}{b} + C_n x_1 \sinh \frac{n\pi x_1}{b} \right.$$

$$\left. + D_n x_1 \cosh \frac{n\pi x_1}{b} \right) \sin \frac{n\pi y}{b} + \sum_{n=1,3,5,...}^{\infty} \frac{4p_0 b^4}{n^5 \pi^5 D} \sin \frac{n\pi y}{b} \qquad (5.9)$$

$$w_2 = \sum_{n=1,3,5,...}^{\infty} \left(K_n \sinh \frac{n\pi x_2}{b} + L_n \cosh \frac{n\pi x_2}{b} + N_n x_2 \sinh \frac{n\pi x_2}{b} \right.$$

$$\left. + Q_n x_2 \cosh \frac{n\pi x_2}{b} \right) \sin \frac{n\pi y}{b} \qquad (5.10)$$

We recall that the boundary conditions along the edges at $y = 0$ and $y = b$ are automatically satisfied by Eqs. (5.9) and (5.10). The unknown constants in Eq. (5.9) are determined by applying the following boundary conditions.

$$w_1 \Big|_{x_1 = 0} = 0 \qquad (5.11a)$$

$$\frac{\partial^2 w_1}{\partial x_1^2} \Big|_{x_1 = 0} = 0 \qquad (5.11b)$$

$$-D \left(\frac{\partial^2 w_1}{\partial x_1^2} + \nu \frac{\partial^2 w_1}{\partial y^2} \right)_{x_1 = a} = \sum_{n=1,3,5,...}^{\infty} M_n \sin \frac{n\pi y}{b} \qquad (5.11c)$$

$$-D \left[\frac{\partial^3 w_1}{\partial x_1^3} + (2 - \nu) \frac{\partial^3 w_1}{\partial y \, \partial x_1^2} \right]_{x_1 = a} = \sum_{n=1,3,5,...}^{\infty} V_n \sin \frac{n\pi y}{b} \qquad (5.11d)$$

From these boundary conditions we obtain four algebraic equations with which we determine A_n, B_n, C_n, and D_n as functions of plate geometry, the material properties, and the constants M_n and V_n. In precisely the same manner, we determine the constants in Eq. (5.10), as functions of M_n and V'_n. Now we must evaluate the three unknowns M_n, V_n, and V'_n. Thus, three more conditions are needed to determine these remaining unknowns. Two conditions are obtained by requiring continuity of geometry along the common edge. The first condition specifies continuity of slope and is the same as we obtained in Section 5.2.

$$\frac{\partial w_1}{\partial x_1} \Big|_{x_1 = 0} = \frac{\partial w_2}{\partial x_2} \Big|_{x_2 = 0} = 0 \qquad (5.12)$$

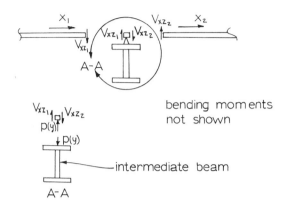

Fig. 5.7.   Shear load carried by supporting beam

The second condition of continuity requires that

$$w_1 \Big|_{x_1 = a} = w_2 \Big|_{x_2 = 0}. \tag{5.13}$$

A third equation containing the unknowns is determined by examining the shear along the common edge. From Fig. 5.7 we see that the difference in shear between Plates 1 and 2 is the load which is carried by the intermediate beam. All forces shown are positive according to our sign convention.

From section A-A in Fig. 5.7 we see that equilibrium requires

$$P(y) = -V_{xz_1} + V_{xz_2} \tag{5.14}$$

in which $P(y)$ is the load on the intermediate beam as shown in Fig. 5.8.

We recall that the differential equation describing the deflection of the neutral axis of a beam is

$$P(y) = EI \, \frac{d^4 w_b}{dy^4}. \tag{5.15}$$

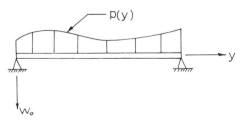

Fig. 5.8.   Loading of intermediate beam

The quantity E is the modulus of elasticity of the material of the beam, and I is the moment of inertia of the beam cross section about its neutral axis. The condition for continuity of shear is obtained by equating Eqs. (5.14) and (5.15).

$$-V_{xz_1} + V_{xz_2} = EI \frac{d^4 w_b}{dy^4} \qquad (5.16)$$

When we apply the condition given by Eq. (6.16) [5.16], we may substitute either $w_b = w_1$ and $x_1 = a$ or $w_b = w_2$ and $x_2 = 0$ into the term $EI\,(d^4 w_b/dy^4)$. If we use w_1, Eq. (5.16) becomes

$$-D\left[-\frac{\partial^3 w_1}{\partial x_1^3} + \frac{\partial^3 w_2}{\partial x_2^3} + (2 - \nu)\left(-\frac{\partial^3 w_1}{\partial x_1\, \partial y^2} + \frac{\partial^3 w_2}{\partial x_2\, \partial y^2} \right) \right]_{\substack{x_1 = a \\ x_2 = 0}}$$

$$= EI \frac{d^4 w_1}{dy^4}\Bigg|_{x_1 = a}. \qquad (5.17)$$

The solution is completed by determining M_n, V_n, and V_n' from Eqs. (5.12), (5.13), and (5.17).

5.4 PLATES WITH INTERMEDIATE ELASTIC SUPPORTS WHICH RESIST ROTATIONS AND DEFLECTIONS

A third variation of the continuous plate occurs when the intermediate elastic support is an integral part of the plate, as shown in Fig. 5.9.

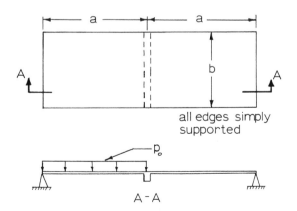

Fig. 5.9. Continuous plate with an integral stiffener

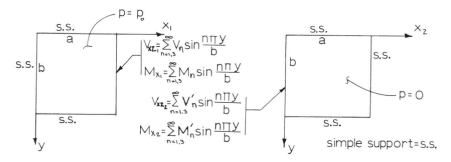

a. Component plate 1   b. Component plate 2

Fig. 5.10. Components of a continuous plate with an integral stiffener

In this case, the intermediate beam resists rotation as well as deflection, and we shall assume that it is restricted against rotation at its ends. We begin the solution by subdividing the plate as shown in Fig. 5.10. The intermediate beam is capable of carrying both shear and moment; thus, the shear and moment along the edge $x_1 = a$ of Plate 1 is not equal to the shear and moment along the edge $x_2 = 0$ of Plate 2. Once again, the general expressions for the lateral deflections of each component plate are described by Eqs. (5.1) and (5.5). If we use the same procedure we used in Sections 5.2 and 5.3, the deflection w_1 may be determined as a function of V_n and M_n, and w_2 may be determined as a function of V'_n and M'_n. Now we must specify four boundary conditions to determine these four unknown constants. Two of these boundary conditions, which come from continuity of slope and deflection along the common edge, are

$$w_1 \Big|_{x_1 = a} = w_2 \Big|_{x_2 = 0} \qquad (5.18)$$

and

$$\frac{\partial w_1}{\partial x_1} \Big|_{x_1 = a} = \frac{\partial w_2}{\partial x_2} \Big|_{x_2 = 0}. \qquad (5.19)$$

Before we can obtain the remaining two boundary conditions, we must investigate the loads on the intermediate beam and on a section of plate adjacent to the intermediate beam. Continuously distributed torsional and lateral loads are transmitted to the intermediate beam from the plate, as shown in Fig. 5.11. Since the beam is in equilibrium, the expression for the distributed torsional load per unit of length, M_b, must be

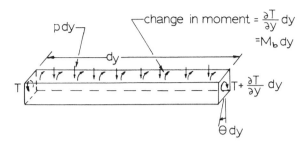

Fig. 5.11. Loading condition of the intermediate beam

$$M_b = \frac{\partial T}{\partial y}. \tag{5.20}$$

We recall that the differential equation describing the angle of twist of a bar subjected to a torsional load is

$$T = JG \frac{\partial \theta}{\partial y} \tag{5.21}$$

where
  θ = angle of twist per unit length
  J = torsional constant
  G = modulus of rigidity.
If we substitute the expression for the torque from Eq. (5.21) into (5.20), the distributed torsional load on the intermediate beam becomes

$$M_b = JG \frac{\partial^2 \theta}{\partial y^2}. \tag{5.22}$$

Since the intermediate beam is rigidly fixed to the plate, the angle of twist of the beam is equal to the slope of the plate as shown in Fig. 5.12.

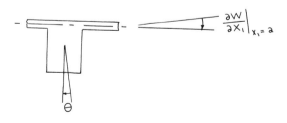

Fig. 5.12. Angle of twist of intermediate beam

Thus, the expression for the angle of twist in terms of the plate deflection is

$$\theta = \frac{\partial \mathbf{w}_1}{\partial \mathbf{x}_1} \Bigg|_{\mathbf{x}_1 = a} . \tag{5.23}$$

Substitution of the expression for the angle of twist from Eq. (5.23) into (5.22) yields the following expression for the distributed torsional load in terms of the plate deflection.

$$\mathbf{M}_b = JG \frac{\partial^3 \mathbf{w}_1}{\partial \mathbf{x}_1 \partial \mathbf{y}^2} \Bigg|_{\mathbf{x}_1 = a} \tag{5.24}$$

The loads acting on a section of the plate of length dy adjacent to the intermediate beam are shown in Fig. 5.13. Since the plate is in

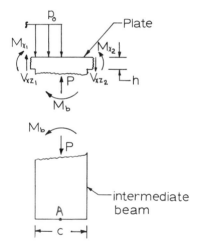

Fig. 5.13. Loads on plate and intermediate beam

equilibrium, the resultant force on the plate section must be zero; thus

$$P + V_{xz_1} - V_{xz_2} - p_0 \frac{c}{2} = 0. \tag{5.25a}$$

The last term of Eq. (5.25a) is very small compared with the other terms, since the dimension c is small compared with the overall plate dimensions; thus, Eq. (5.25a) becomes

$$P = -V_{xz_1} + V_{xz_2} \tag{5.25b}$$

or

$$EI \left. \frac{d^4 w_1}{dy^4} \right|_{x_1 = a} = D \left[\frac{\partial^3 w_1}{\partial x_1^3} + (2 - \nu) \frac{\partial^3 w_1}{\partial x_1 \partial y^2} \right]_{x_1 = a}$$
$$- D \left[\frac{\partial^3 w_2}{\partial x_2^3} + (2 - \nu) \frac{\partial^3 w_2}{\partial x_2 \partial y^2} \right]_{x_2 = 0} \tag{5.25c}$$

which is identical to Eq. (5.17).

For equilibrium, the resultant moment on the plate section also must be zero. If we determine the expression for the moment about an axis parallel to the y axis and passing through point A of the plate section in Fig. 5.13, we have

$$M_b + M_{x_1} - M_{x_2} + \frac{c}{2} V_{xz_1} + \frac{c}{2} V_{xz_2} - \frac{c}{4} p_0 \frac{c}{2} = 0. \tag{5.26a}$$

The last three terms of Eq. (5.26a) are very small compared with the other terms, since the dimension c is small compared with the overall plate dimensions. Thus, Eq. (5.26a) becomes

$$M_b = -M_{x_1} + M_{x_2}. \tag{5.26b}$$

Substitution of Eq. (5.24) into (5.26b) yields

$$JG \left. \frac{\partial^3 w_1}{\partial x_1 \partial y^2} \right|_{x_1 = a} = -M_{x_1} + M_{x_2} \tag{5.26c}$$

and, if we expand the terms on the right-hand side, we obtain

$$JG \left. \frac{\partial^3 w_1}{\partial x_1 \partial y^2} \right|_{x_1 = a} = D \left[\frac{\partial^2 w_1}{\partial x_1^2} + \nu \frac{\partial^2 w_1}{\partial y^2} \right]_{x_1 = a}$$
$$- D \left[\frac{\partial^2 w_2}{\partial x_2^2} + \nu \frac{\partial^2 w_2}{\partial y^2} \right]_{x_2 = 0}. \tag{5.26d}$$

The four unknown constants V_n, M_n, V'_n, and M'_n can be determined from the four algebraic equations obtained by substituting the expressions for w_1 and w_2 into the conditions given by Eqs. (5.18), (5.19), (5.25c), and (5.26d), and the solution is completed.

5.5 PLATES SUPPORTED BY INTERMEDIATE COLUMNS

Frequently in structural analysis, we encounter thin plates with intermediate column supports. In this section we will consider plates with a limited number of intermediate columns. We will not attempt to specifically define "limited number" except to say that each new column adds an additional algebraic equation to the set of simultaneous equations required for the solution. Consequently, the real limits will be dependent upon the computer facilities available and the desire of the person solving the problem. In Chapter 7 we will find that certain problems with intermediate columns can be solved by the method of Ritz with Lagrange multipliers.

Our procedure here is more easily presented through the use of an example problem. For this example, let us consider the simply supported plate with the concentrated loads in Fig. 5.14. It should be mentioned that the following procedure will apply in general to plates of arbitrary loading and boundary conditions.

The column reactions and lateral deflections can be solved by utilizing Navier solutions. For a first approximation let us assume that column cross sections are so small that their reactions on the plate are considered to be point loads, as shown in Fig. 5.15, and that the columns do not restrain the plate from rotation. Thus, a column reaction is only a vertical load. The plate in Fig. 5.14 has a relatively simple solution

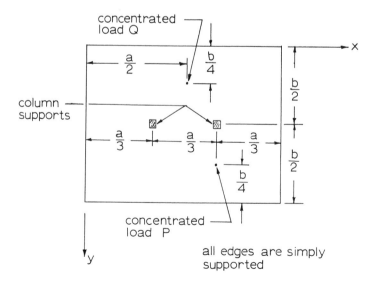

Fig. 5.14. Simply supported plate with intermediate columns

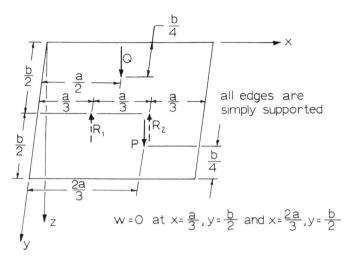

Fig. 5.15. Simply supported plates with concentrated loads and column reactions

because it has only two columns. However, we will see that precisely the same technique can be extended to a plate with additional columns.

The expression for the lateral deflection corresponding to the plate of Fig. 5.15 is obtained from Eq. (3.17c) as follows.

$$w = \sum_{i=1}^{4} K(x,y; \zeta_i, \eta_i) P_i(\zeta_i, \eta_i) \qquad (5.27a)$$

or

$$w = QK(x,y; a/2, b/4) + PK(x,y; 2a/3, 3b/4)$$
$$- R_1 K(x,y; a/3, b/2) - R_2(x,y; 2a/3, b/2) \qquad (5.27b)$$
$$-R_2 K$$

The expression for K may be obtained from Eq. (3.17a); thus, Eq. (5.27b) becomes, after simplification

$$w = \sum_{m=1}^{\infty} \sum_{n=1}^{\infty} \frac{4}{ab\pi^4 D \left[\left(\dfrac{m}{a} \right)^2 + \left(\dfrac{n}{b} \right)^2 \right]^2} \left(Q \sin \frac{m\pi}{2} \sin \frac{n\pi}{4} \right.$$
$$+ P \sin \frac{2m\pi}{3} \sin \frac{3n\pi}{4} - R_1 \sin \frac{m\pi}{3} \sin \frac{n\pi}{2}$$
$$\left. - R_2 \sin \frac{2m\pi}{3} \sin \frac{n\pi}{2} \right) \sin \frac{m\pi x}{a} \sin \frac{n\pi y}{b} . \qquad (5.27c)$$

Again it should be pointed out that an expression for the deflection, such as that of Eq. (5.27c), can be obtained for any type of loading or boundary conditions by referring to the appropriate solutions in Chapter 3. The column reactions are obtained from the conditions

$$w \bigg|_{\substack{x \,=\, a/3 \\ y \,=\, b/2}} = 0 \qquad\qquad (5.28a)$$

$$w \bigg|_{\substack{x \,=\, 2a/3 \\ y \,=\, b/2}} = 0. \qquad\qquad (5.28b)$$

These two conditions lead to two linear algebraic equations which can be solved for the unknown column reactions R_1 and R_2. For the convenience of notation we define

$$C_{mn} = \frac{4}{ab\pi D^4 \left[\left(\dfrac{m}{a}\right)^2 + \left(\dfrac{n}{b}\right)^2 \right]^2} \sin\frac{m\pi}{3} \sin\frac{n\pi}{2}$$

$$C_1 = \sum_{m=1}^{\infty} \sum_{n=1}^{\infty} C_{mn} \sin\frac{m\pi}{2} \sin\frac{n\pi}{4}$$

$$C_2 = \sum_{m=1}^{\infty} \sum_{n=1}^{\infty} C_{mn} \sin\frac{2m\pi}{3} \sin\frac{3n\pi}{4}$$

$$C_3 = \sum_{m=1}^{\infty} \sum_{n=1}^{\infty} C_{mn} \sin\frac{m\pi}{3} \sin\frac{m\pi}{2}$$

$$C_4 = \sum_{m=1}^{\infty} \sum_{n=1}^{\infty} C_{mn} \sin\frac{2m\pi}{3} \sin\frac{n\pi}{2}$$

$$K_{mn} = \frac{4}{ab\pi D^4 \left[\left(\dfrac{m}{a}\right)^2 + \left(\dfrac{n}{b}\right)^2 \right]^2} \sin\frac{2m\pi}{3} \sin\frac{n\pi}{2}$$

$$C_5 = \sum_{m=1}^{\infty} \sum_{n=1}^{\infty} K_{mn} \sin\frac{m\pi}{2} \sin\frac{n\pi}{4}$$

$$C_6 = \sum_{m=1}^{\infty} \sum_{n=1}^{\infty} K_{mn} \sin\frac{2m\pi}{3} \sin\frac{3n\pi}{4}$$

$$C_7 = \sum_{m=1}^{\infty} \sum_{n=1}^{\infty} K_{mn} \sin \frac{m\pi}{3} \sin \frac{n\pi}{2}$$

$$C_8 = \sum_{m=1}^{\infty} \sum_{n=1}^{\infty} K_{mn} \sin \frac{2m\pi}{3} \sin \frac{n\pi}{2}.$$

If we apply the conditions given by Eqs. (5.28a) and (5.28b) to Eq. (5.27c), we obtain

$$C_1 Q + C_2 P - C_3 R_1 - C_4 R_2 = 0 \qquad (5.29a)$$

$$C_5 Q + C_6 P - C_7 R_1 - C_8 R_2 = 0. \qquad (5.29b)$$

The simultaneous solution of Eqs. (5.29a) and (5.29b) gives

$$R_1 = \frac{C_1 C_8 - C_4 C_5}{C_3 C_8 - C_4 C_7} Q + \frac{C_2 C_8 - C_4 C_6}{C_3 C_8 - C_4 C_7} P \qquad (5.30a)$$

and

$$R_2 = \frac{C_1 C_7 - C_3 C_5}{C_4 C_7 - C_3 C_8} Q + \frac{C_2 C_7 - C_3 C_6}{C_4 C_7 - C_3 C_8} P. \qquad (5.30b)$$

Substituting Eqs. (5.30a) and (5.30b) into Eq. (5.27c) completes the solution for the lateral deflection. It is evident that each time we add an additional column, we increase the number of equations to be solved simultaneously by one. We evaluate $w = 0$ at each new column, thereby adding an additional algebraic equation with which we can solve for the new unknown reaction. Consequently, in theory this technique applies equally well if many intermediate columns exist. In practice the algebra rapidly becomes unwieldy as additional columns are added.

It is possible that we may want to improve our model by assuming that the column reaction is uniformly distributed over a small rectangle, instead of a point, as shown in Fig. 5.16. The expression for the lateral deflection corresponding to the plate in Fig. 5.16 is obtained from Eqs. (3.10) and (3.16) and the principle of superposition as follows.

$$w = \sum_{m=1}^{\infty} \sum_{n=1}^{\infty} \frac{4}{ab\pi^4 D \left[\left(\frac{m}{a}\right)^2 + \left(\frac{n}{b}\right)^2 \right]^2} \left(Q \sin \frac{m\pi}{2} \sin \frac{n\pi}{4} \right.$$

$$\left. + P \sin \frac{2m\pi}{3} \sin \frac{3n\pi}{4} \right) \sin \frac{m\pi x}{a} \sin \frac{n\pi y}{b}$$

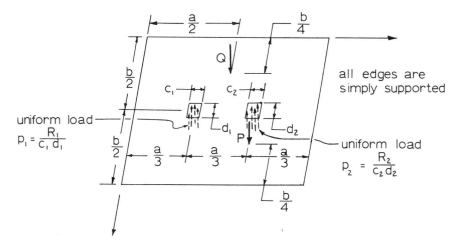

Fig. 5.16. Plate with uniformly loaded column reactions

$$
- \sum_{m=1}^{\infty} \sum_{n=1}^{\infty} \frac{16}{mn\pi^6 D \left[\left(\dfrac{m}{a} \right)^2 + \left(\dfrac{n}{b} \right)^2 \right]^2}
$$
$$
\left(q_1 \sin \frac{m\pi}{3} \sin \frac{m\pi c_1}{2a} \sin \frac{n\pi}{2} \sin \frac{n\pi d_1}{2b} \right.
$$
$$
\left. + q_2 \sin \frac{2m\pi}{3} \sin \frac{m\pi c_2}{2a} \sin \frac{n\pi}{2} \sin \frac{n\pi d_2}{2b} \right) \sin \frac{m\pi x}{a} \sin \frac{n\pi y}{b} \quad (5.31)
$$

If we require once again that $w = 0$ at the center point of each column, we obtain two linear algebraic equations which can be solved for the unknown column reactions q_1 and q_2.

If we use Eqs. (1.11), we find that the moments near the columns are much larger than the moments some distance away from the columns.[1] We also find that these moments are largely dependent upon the cross-sectional dimensions of the columns. In contrast, the moments halfway between the columns are nearly independent of the dimensions of the column cross section.

Again we repeat that the previous technique can be used regardless of the type of applied loads or boundary conditions. The general procedure for determining the deflection of a plate with column supports is summarized in the following steps.

1. Determine the expression for the deflection of the plate for each applied load and each column reaction. Each of these expressions comes from either the Navier solution (simply supported edges), the Levy solution (two edges simply supported and the other two arbitrarily sup-

ported), or the Levy solution with the principle of superposition (other boundary conditions).

2. Add the expressions of Step 1 above to obtain the expression for the total deflection. This expression will be in terms of the unknown column reactions.

3. Require that the deflection be zero at the center of each column reaction. This requirement leads to a set of linear equations which can be solved simultaneously for the column reactions.

4. Substitute the column reactions obtained in Step 3 into the expression for the deflection of Step 2, and the solution is completed.

If the plate has (1) a uniform load, (2) equally spaced columns, and (3) overall dimensions that are large compared with the column spacing, the symmetry of geometry and loading may be utilized to obtain a simplified expression for the lateral deflection.[1]

5.6 PLATES ON ELASTIC FOUNDATIONS

Often a plate is placed on a continuous elastic foundation. Examples of this type of plate are streets, airport runways, and slab foundations for buildings. In the analyses of these plates it is assumed that the intensity of the reaction at any point is proportional to the deflection at that point. In essence, we assume that the elastic foundation is a large continuous spring with the relationship

$$F = k \Delta \tag{5.32}$$

where
  F = foundation force/unit of foundation area
  Δ = foundation deflection
  k = foundation stiffness constant =
  $\dfrac{\text{foundation force/unit of foundation area}}{\text{unit of foundation deflection}}$.

The constant k has a wide range of values, and can best be determined through bearing tests of the actual foundation material. However, if the plate is large and bearing tests are not possible, the average values of k given in Table 3 serve as good approximations.

Simply Supported Plate. We shall determine the lateral deflection of a simply supported plate on an elastic foundation as shown in Fig. 5.17.

The lateral load acting on the plate is $p(x,y)$ and the reaction of the elastic foundation is $F(x,y)$; thus, the resultant distributed load is

TABLE 3
Foundation Stiffness Constants[22]

Major Divisions	Soil Groups and Typical Description	Approximate Range of k-values, psi/in
Gravel and gravelly soils	Well-graded gravel and gravel-sand mixtures; little or no fines	500–700 or greater
	Well-graded gravel-sand-clay mixtures; excellent binder	400–700 or greater
	Poorly graded gravel and gravel-sand mixtures; little or no fines	300–500
	Gravel with fines, very silty gravel, clayey gravel, poorly graded gravel-sand-clay mixtures	250–500
Sands and sandy soils	Well-graded sands and gravelly sands; little or no fines	250–575
	Well-graded sand-clay mixtures; excellent binder	250–575
	Poorly graded sands; little or no fines	200–325
	Sands with fines, very silty sands, clayey sands, poorly graded sand-clay mixtures	175–325
Fine-grained soils having low to medium compressibility	Silts (inorganic) and very fine sands, rock flour, silty or clayey fine sands with slight plasticity	150–300
	Clays (inorganic) of low to medium plasticity, sandy clays, silty clays, lean clays	125–225
	Organic silts and organic silt-clays of low plasticity	100–175
Fine-grained soils having high compressibility	Micaceous or diatomaceous fine sandy and silty soils, elastic silts	50–175
	Clays (inorganic) of high plasticity, fat clays	50–150
	Organic clays of medium to high plasticity	50–125

$$p'(x,y) = p(x,y) - F(x,y)$$

as shown in Fig. 5.17. The partial differential equation of equilibrium becomes

$$\nabla^4 w(x,y) = \frac{p'(x,y)}{D} = \frac{p(x,y) - F(x,y)}{D} = \frac{p(x,y) - kw(x,y)}{D}. \quad (5.33)$$

Since the plate is simply supported on all four edges, we may determine the deflection by the method of Navier. According to Eq. (3.1), we assume the deflection to be

$$w = \sum_{m=1}^{\infty} \sum_{n=1}^{\infty} W_{mn} \sin \frac{m\pi x}{a} \sin \frac{n\pi y}{b}. \quad (5.34)$$

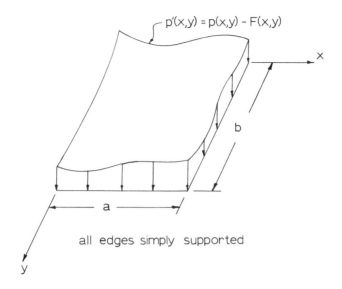

Fig. 5.17. Loading of a rectangular plate on an elastic foundation

The foundation force becomes

$$F = kw = k \sum_{m=1}^{\infty} \sum_{n=1}^{\infty} W_{mn} \sin \frac{m\pi x}{a} \sin \frac{n\pi y}{b}. \qquad (5.35)$$

According to Eq. (3.3), we express the known lateral load distribution in the form

$$p(x,y) = \sum_{m=1}^{\infty} \sum_{n=1}^{\infty} A_{mn} \sin \frac{m\pi x}{a} \sin \frac{n\pi y}{b} \qquad (5.36)$$

where

$$A_{mn} = \frac{4}{ab} \int_0^a \int_0^b p(x,y) \sin \frac{m\pi x}{a} \sin \frac{n\pi y}{b} \, dxdy. \qquad (5.37)$$

Substitution of Eqs. (5.34), (5.35), and (5.36) into Eq. (5.33) yields

$$\sum_{m=1}^{\infty} \sum_{n=1}^{\infty} W_{mn} \left(\frac{m^4 \pi^4}{a^4} + 2 \frac{m^2 n^2 \pi^4}{a^2 b^2} + \frac{n^4 \pi^4}{b^4} \right) \sin \frac{m\pi x}{a} \sin \frac{n\pi y}{b}$$

$$= \frac{1}{D} \sum_{m=1}^{\infty} \sum_{n=1}^{\infty} (A_{mn} - kW_{mn}) \sin \frac{m\pi x}{a} \sin \frac{n\pi y}{b}$$

or

$$\sum_{m=1}^{\infty} \sum_{n=1}^{\infty} \left\{ \pi^4 \left[\left(\frac{m}{a}\right)^2 + \left(\frac{n}{b}\right)^2 \right]^2 W_{mn} \right.$$
$$\left. - \frac{1}{D} (A_{mn} - kW_{mn}) \right\} \sin \frac{m\pi x}{a} \sin \frac{n\pi y}{b} = 0. \qquad (5.38)$$

Equation (5.38) must be satisfied for all values of x and y; thus, we have

$$\pi^4 \left[\left(\frac{m}{a}\right)^2 + \left(\frac{n}{b}\right)^2 \right]^2 W_{mn} - \frac{1}{D} (A_{mn} - kW_{mn}) = 0$$

or

$$W_{mn} = \frac{A_{mn}}{\pi^4 D \left[\left(\frac{m}{a}\right)^2 + \left(\frac{n}{b}\right)^2 \right]^2 + k}. \qquad (5.39)$$

The deflection becomes

$$w = \sum_{m=1}^{\infty} \sum_{n=1}^{\infty} \frac{A_{mn}}{\pi^4 D \left[\left(\frac{m}{a}\right)^2 + \left(\frac{n}{b}\right)^2 \right]^2 + k} \sin \frac{m\pi x}{a} \sin \frac{n\pi y}{b} \qquad (5.40)$$

where A_{mn} is determined from Eq. (5.37) for any arbitrary loading.

If the plate has a uniformly distributed load of intensity p_0 acting over a small area as shown in Fig. 3.2, the expression for A_{mn} is

$$A_{mn} = \frac{4}{ab} \int_{\eta_0 - d/2}^{\eta_0 + d/2} \int_{\zeta_0 - c/2}^{\zeta_0 + c/2} p_0 \sin \frac{m\pi x}{a} \sin \frac{n\pi y}{b} \, dx \, dy. \qquad (5.41)$$

After integration, Eq. (5.41) becomes

$$A_{mn} = \frac{16 p_0}{mn\pi^2} \sin \frac{m\pi \zeta_0}{a} \sin \frac{m\pi c}{2a} \sin \frac{n\pi \eta_0}{b} \sin \frac{n\pi d}{2b} \qquad (5.42)$$

and the expression for the deflection is obtained by substituting the value of A_{mn} from Eq. (5.42) into (5.40).

If the plate has a concentrated load P at (ζ_0, η_0), we obtain the expression for the deflection by defining the total load on the area cd as

$$P = p_0 \, cd \qquad (5.43a)$$

from which we obtain an expression for p_0 as

$$p_0 = \frac{P}{cd}. \qquad (5.43b)$$

Substitution of Eq. (5.43b) into (5.42) yields

$$A_{mn} = \frac{16P}{mn\pi^2 cd} \sin \frac{m\pi \zeta_0}{a} \sin \frac{m\pi c}{2a} \sin \frac{n\pi \eta_0}{b} \sin \frac{n\pi d}{2b}. \qquad (5.44)$$

If c and d are made to approach zero while we let p_0 approach infinity such that the product defined by Eq. (5.43a) remains constant, then in the limit the load becomes a concentrated load of magnitude P acting at the point (ζ_0, η_0) and Eq. (5.42) becomes

$$A_{mn} = \lim_{\substack{c \to 0 \\ d \to 0}} \left(\frac{16P}{mn\pi^2 cd} \sin \frac{m\pi \zeta_0}{a} \sin \frac{m\pi c}{2a} \sin \frac{n\pi \eta_0}{b} \sin \frac{n\pi d}{2b} \right)$$

$$= \left(\frac{16P}{mn\pi^2} \sin \frac{m\pi \zeta_0}{a} \sin \frac{n\pi \eta_0}{b} \right) \left(\lim_{c \to 0} \frac{\sin \frac{m\pi c}{2a}}{c} \right) \left(\lim_{d \to 0} \frac{\sin \frac{n\pi d}{2b}}{d} \right)$$

$$= \left(\frac{16P}{mn\pi^2} \sin \frac{m\pi \zeta_0}{a} \sin \frac{n\pi \eta_0}{b} \right) \left(\frac{m\pi}{2a} \right) \left(\frac{n\pi}{2b} \right)$$

$$= \frac{4P}{ab} \sin \frac{m\pi \zeta_0}{a} \sin \frac{n\pi \eta_0}{b} \qquad (5.45)$$

and the expression for the deflection once again is obtained by substituting the value for A_{mn} into Eq. (5.40).

Plates with Arbitrary Boundary Conditions. If the plate has two opposite edges simply supported, we may determine the lateral deflection by a modified Levy method. We shall present an example of this method by considering the plate in Fig. 5.18.

The boundary conditions of the plate in Fig. 5.18 are

$$w \Big|_{x = 0,a} = 0 \qquad (5.46a)$$

$$\frac{\partial^2 w}{\partial x^2} \Bigg|_{x = 0,a} = 0 \qquad (5.46b)$$

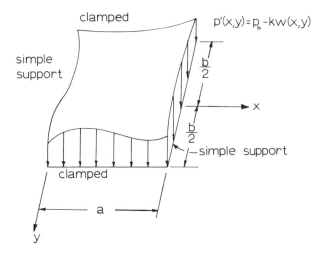

Fig. 5.18.   Uniformly loaded plate on an elastic foundation

$$w\bigg|_{y\,=\,\frac{b}{2},\,-\,\frac{b}{2}} = 0 \qquad (5.46c)$$

$$\frac{\partial w}{\partial y}\bigg|_{y\,=\,\frac{b}{2},\,-\,\frac{b}{2}} = 0. \qquad (5.46d)$$

The partial differential equation of equilibrium is again

$$\nabla^4 w = \frac{p_0 - kw}{D}$$

or

$$\nabla^4 w + \frac{k}{D}\,w = \frac{p_0}{D}. \qquad (5.47)$$

According to Eq. (3.18) the solution is assumed to consist of a homogeneous part and a particular part

$$w = w_h + w_p \qquad (5.48)$$

in which w_h and w_p represent the homogeneous and particular solutions respectively. This assumption leads to the equation

$$\nabla^4 w_h + \frac{k}{D} w_h = 0 \qquad (5.49a)$$

from which the homogeneous solution is determined, and to the equation

$$\nabla^4 w_p + \frac{k}{D} w_p = \frac{p_0}{D} \qquad (5.49b)$$

from which the particular solution is determined. The homogeneous solution is assumed in the form

$$w_h = \sum_{m=1,3,5,\ldots}^{\infty} f_m(y) \sin \frac{m\pi x}{a} \qquad (5.50)$$

where $f_m(y)$ is an arbitrary function of y. Note that we have eliminated the even values of n from Eq. (5.50) since the applied loading and boundary conditions are symmetric with respect to the line $x = a/2$. We also see that Eq. (5.50) satisfies the boundary conditions given by Eqs. (5.46a) and (5.46b). If we substitute Eq. (5.50) into (5.49a), we obtain

$$\sum_{m=1,3,5,\ldots}^{\infty} \left(\frac{d^4 f_m}{dy^4} - 2 \frac{m^2 \pi^2}{a^2} \frac{d^2 f_m}{dy^2} + \frac{m^4 \pi^4}{a^4} f_m + \frac{k}{D} f_m \right) \sin \frac{m\pi x}{a} = 0. \qquad (5.51)$$

Equation (5.51) is satisfied for all values of x if

$$\frac{d^4 f_m}{dy^4} - 2 \frac{m^2 \pi^2}{a^2} \frac{d^2 f_m}{dy^2} + \left(\frac{m^4 \pi^4}{a^4} + \frac{k}{D} \right) f_m = 0. \qquad (5.52)$$

We recall that a linear differential equation with constant coefficients has a solution of the form

$$f_m = E_m e^{\lambda_m y}. \qquad (5.53)$$

If we substitute Eq. (5.53) into (5.52), we obtain the characteristic equation

$$\lambda_m^4 - 2 \frac{m^2 \pi^2}{a^2} \lambda_m^2 + \left(\frac{m^4 \pi^4}{a^4} + \frac{k}{D} \right) = 0. \qquad (5.54)$$

The roots of this characteristic equation can easily be determined as

$$\lambda_{m1} = A_m + iB_m \qquad (5.55a)$$

$$\lambda_{m2} = A_m - iB_m \qquad (5.55b)$$

$$\lambda_{m3} = -A_m + iB_m \qquad (5.55c)$$

$$\lambda_{m4} = -A_m - iB_m \qquad (5.55d)$$

where

$$A_m = \left[\frac{1}{2} \left(\frac{m^2 \pi^2}{a^2} + \sqrt{\frac{m^4 \pi^4}{a^4} + \frac{k}{D}} \right) \right]^{1/2}$$

$$B_m = \left[\frac{1}{2} \left(\frac{m^2 \pi^2}{a^2} - \sqrt{\frac{m^4 \pi^4}{a^4} + \frac{k}{D}} \right) \right]^{1/2}.$$

The expression for f_m becomes

$$f_m = E_{1m} e^{(A_m + iB_m)y} + E_{2m} e^{(A_m - iB_m)y} + E_{3m} e^{(-A_m + iB_m)y} + E_{4m} e^{(-A_m - iB_m)y}$$

or

$$f_m = E'_{1m} \sinh A_m y \sin B_m y + E'_{2m} \sinh A_m y \cos B_m y$$
$$+ E'_{3m} \cosh A_m y \sin B_m y + E'_{4m} \cosh A_m y \cos B_m y. \qquad (5.56)$$

Because of the symmetry of boundary conditions and loading about the x axis, f_m must be an even function of y. Thus

$$E'_{2m} = E'_{3m} = 0$$

and the expression for the particular solution becomes

$$w_h = \sum_{m=1,3,5,\ldots}^{\infty} (E'_{1m} \sinh A_m y \sin B_m y$$
$$+ E'_{4m} \cosh A_m y \cos B_m y) \sin \frac{m \pi x}{a}. \qquad (5.57)$$

For the particular solution, we first express the load p_0 in terms of a Fourier sine series, according to Eq. 3.28.

$$p_0 = \sum_{m=1}^{\infty} p_m(y) \sin \frac{m \pi x}{a} \qquad (5.58)$$

where

$$p_m(y) = \frac{2}{a} \int_0^a p_0 \sin \frac{m\pi x}{a} \, dx$$

$$= \begin{cases} \dfrac{4p_0}{m\pi} & \text{for} \quad m = 1, 3, 5, \ldots \\ 0 & \text{for} \quad m = 2, 4, 6, \ldots \end{cases} \tag{5.59}$$

Next, we assume w_p in the form

$$w_p = \sum_{m=1}^{\infty} W_m(y) \sin \frac{m\pi x}{a} \tag{5.60}$$

where $W_m(y)$ is an arbitrary function of y. We see that w_p also satisfies the boundary conditions given by Eqs. (5.46a) and (5.46b). If we substitute Eqs. (5.59) and (5.60) into Eq. (5.49b), we have

$$\sum_{m=1,3,5,\ldots}^{\infty} \left(\frac{d^4 W_{mn}}{dy^4} - 2 \frac{m^2 \pi^2}{a^2} \frac{d^2 W_{mn}}{dy^2} + \frac{m^4 \pi^4}{a^4} W_{mn} + \frac{k}{D} W_{mn} \right) \sin \frac{m\pi x}{a}$$

$$= \sum_{m=1,3,5,\ldots}^{\infty} \frac{4p_0}{Dm\pi} \sin \frac{m\pi x}{a}. \tag{5.61}$$

Equation (5.61) is satisfied for all values of x if

$$\frac{d^4 W_{mn}}{dy^4} - 2 \frac{m^2 \pi^2}{a^2} \frac{d^2 W_{mn}}{dy^2} + \left(\frac{m^4 \pi^4}{a^4} + \frac{k}{D} \right) W_{mn} = \frac{4p_0}{Dm\pi}. \tag{5.62}$$

The particular solution of Eq. (5.62) is

$$W_{mn} = \frac{4p_0}{Dm\pi \left(\dfrac{m^4 \pi^4}{a^4} + \dfrac{k}{D} \right)}. \tag{5.63}$$

Thus, the final expression for w_p is

$$w_p = \frac{4p_0}{\pi D} \sum_{m=1,3,5,\ldots}^{\infty} \frac{1}{m \left(\dfrac{m^4 \pi^4}{a^4} + \dfrac{k}{D} \right)} \sin \frac{m\pi x}{a} \tag{5.64}$$

and the general expression for the solution becomes

$$\mathbf{w} = \mathbf{w}_h + \mathbf{w}_p$$

$$= \sum_{m=1,3,5,\ldots}^{\infty} \left[E'_{1m} \sinh A_m y \sin B_m y + E'_{4m} \cosh A_m y \cos B_m y \right.$$

$$\left. + \frac{4p_0}{m\pi D \left(\dfrac{m^4 \pi^4}{a^4} + \dfrac{k}{D} \right)} \right] \sin \frac{m\pi x}{a}. \qquad (5.65)$$

The four boundary conditions that remain to be satisfied are the two conditions given in Eq. (5.46c) and the two conditions given in Eq. (5.46d). Because of the previously described symmetry of the problem, we need only to satisfy one of the conditions of Eq. (5.46c) and one of the conditions of Eq. (5.46d). We choose

$$\mathbf{w} \Big|_{y = \frac{b}{2}} = 0 \qquad (5.66a)$$

$$\frac{\partial \mathbf{w}}{\partial y} \Big|_{y = \frac{b}{2}} = 0. \qquad (5.66b)$$

Substitution of Eq. (5.65) into (5.66) yields the two algebraic equations

$$\frac{4p_0}{m\pi D \left(\dfrac{m^4 \pi^4}{a^4} + \dfrac{k}{D} \right)} + E'_{1m} \sinh A_m \frac{b}{2} \sin B_m \frac{b}{2}$$

$$+ E'_{4m} \cosh A_m \frac{b}{2} \cos B_m \frac{b}{2} = 0 \qquad (5.67a)$$

$$E'_{1m} \left(B_m \sinh A_m \frac{b}{2} \cos B_m \frac{b}{2} + A_m \cosh A_m \frac{b}{2} \sin B_m \frac{b}{2} \right)$$

$$+ E'_{4m} \left(-B_m \cosh A_m \frac{b}{2} \sin B_m \frac{b}{2} + A_m \sinh A_m \frac{b}{2} \cos B_m \frac{b}{2} \right) = 0 \quad (5.67b)$$

from which the constants E'_{1m} and E'_{4m} can be solved, and the solution is completed. If the plate has boundary conditions other than that of simple supports on two opposite sides, the modified Levy method cannot be used; however, we may use the method of superposition combined with the modified Levy solution presented in Section 3.3.

Problems

23. Determine an expression for the lateral deflections of the uniformly loaded plate in Fig. 5.19. All exterior plate boundaries are simply supported.

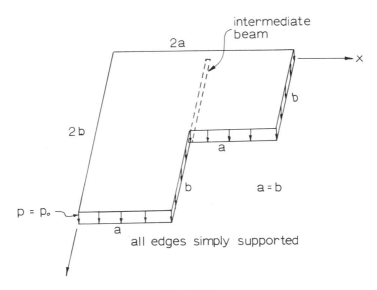

Fig. 5.19

a. Assume the intermediate beam does not resist rotation but is sufficiently rigid that lateral deflection along Line c-c can be considered negligible.

b. Assume the intermediate beam does not resist rotation and is sufficiently flexible to permit significant lateral deflections. Assume the bending stiffness of this beam is a constant EI_0 along the length of the beam.

c. Assume the intermediate beam is attached to the plate in such a manner that it can be considered an integral part of the plate, i.e. the beam carries torsional and bending loads. Assume the torsional and bending stiffnesses of the beam are GJ_0 and EI_0 respectively.

24. The simply supported plate in Fig. 5.20 is also supported by intermediate columns as shown. Assume the columns act as point supports. Calculate the column reactions and the deflection at the center.

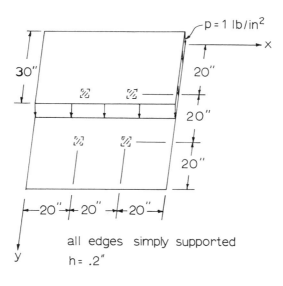

Fig. 5.20

25. Let's assume the columns in Fig. 5.19 are of significant size and can no longer be considered point supports. Let's also assume the cross section of each support is a four inch square. Compute the deflections at (30, 20) and (30, 30).

26. Let's assume the uniformly loaded simply supported plate in Fig. 5.21 rests on an elastic foundation with a spring constant k. Let's also assume that h = .1 in and E = 30 × 10⁶ psi. Determine the necessary spring constant k to keep the maximum deflection less than .05 in.

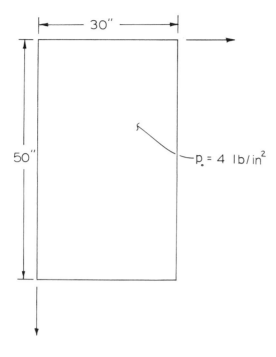

Fig. 5.21

Chapter 6
ORTHOTROPIC PLATES

6.1 INTRODUCTION

All our previous derivations have assumed the plate material to be isotropic according to Assumption 4 in Section 1.1. We recall that the assumption of isotropy implies that the material properties at a point are the same in all directions. However, certain materials have properties that are not independent of the direction; consequently, these materials are said to be anisotropic. One of the most familiar anisotropic materials is wood. The obvious differences in the properties of wood in the directions parallel and perpendicular to the grain are illustrated in Table 4. Other anisotropic materials are concrete and the composite material fiberglass.

TABLE 4
Material Properties of Ash and Birch Woods[22]

| Wood Type | Material properties parallel to the grain are divided by material properties perpendicular to the grain | | |
	$\dfrac{E \text{ (parallel)}}{E \text{ (perpendicular)}}$	$\dfrac{G \text{ (parallel)}}{G \text{ (perpendicular)}}$	$\dfrac{\text{Compression Stress at Proportional Limit Parallel}}{\text{Compression Stress at Proportional Limit Perpendicular}}$
Ash	$9.2 \rightarrow 15.6$	$2.4 \rightarrow 3.6$	4.1
Birch	$12.8 \rightarrow 20$	$3.9 \rightarrow 4.4$	5.2

The general concept of anisotropy is not new, since Cauchy[23] wrote a paper in 1828 giving generalized equations for the elasticity of anisotropic materials. Since that time, many persons have conducted studies of anisotropic materials. A fairly comprehensive account of the theory of elasticity of anisotropic materials is presented by Lechnitsky in two books titled *Anisotropic Plates*[24] and *Theory of Elasticity of an Anisotropic Body*.[25]

In this text we are concerned only with the special case of anisotropy in which the material properties are different in two mutually perpendicular directions. Materials that exhibit this type of behavior are said to be orthogonally anisotropic, or orthotropic. Fabrication methods sometimes make it necessary to consider conditions of orthotropy for structural materials. Some sheets of metal show different elastic properties in different directions, dependent upon the direction of rolling when they are formed. Often, plates such as corrugated plates and plates with stiffeners are mathematically modeled as orthotropic plates.

In this chapter we present the governing equation for the small deflections of the middle surface of thin rectangular orthotropic plates. The governing equation and the equations describing strain, stress, and stress resultants are presented in terms of orthotropic constants. It is shown how structures such as reinforced concrete slabs, plates reinforced by stiffeners and ribs, and corrugated plates can be approximated by orthotropic plates. The appropriate orthotropic constants for each case are presented. A solution is obtained for a simply supported orthotropic plate according to the method of Navier (Chapter 3). The Levy technique (Chapter 3) is used to obtain a solution for an orthotropic plate simply supported on two opposite edges and arbitrarily supported on the other two edges. Additional and more complicated problems involving orthotropic plates are considered in Chapter 9 where the method of finite elements is presented.

6.2   GOVERNING EQUATION

The derivation of the governing differential equation defining small lateral deflections of the middle surface of a thin orthotropic plate is very similar to the derivation presented in Chapter 1 for an isotropic plate. The stress-strain equations given by Eqs. (1.9) are not valid for an orthotropic plate; thus, we must obtain a new set of stress-strain relations that reflect the orthotropic properties of the plate. Such a set of relations is[1]

$$\sigma_x = E_x \epsilon_x + E_{xy} \epsilon_y \tag{6.1a}$$

$$\sigma_y = E_{xy}\,\epsilon_x + E_y\,\epsilon_y \tag{6.1b}$$

$$\tau_{xy} = G\,\gamma_{xy} \tag{6.1c}$$

where the constants E_x, E_{xy}, and E_y characterize the elastic properties of the orthotropic material. Another form of these relationships found in some of the literature is

$$\sigma_x = \frac{E'_x}{1 - \nu_x \nu_y}\,(\epsilon_x + \nu_y\,\epsilon_y) \tag{6.2a}$$

$$\sigma_y = \frac{E'_y}{1 - \nu_x \nu_y}\,(\epsilon_y + \nu_x\,\epsilon_x) \tag{6.2b}$$

$$\tau_{xy} = G\,\gamma_{xy} \tag{6.2c}$$

in which ν_x and ν_y are effective Poisson's ratios in the x and y directions respectively, and E'_x and E'_y are effective moduli of elasticity in the x and y directions respectively. By comparing Eqs. (6.2) with (6.1) we find the relationships between the two sets of elastic constants to be

$$E_x = \frac{E'_x}{1 - \nu_x \nu_y} \tag{6.3a}$$

$$E_y = \frac{E'_y}{1 - \nu_x \nu_y} \tag{6.3b}$$

$$E_{xy} = \frac{E'_x \nu_y}{1 - \nu_x \nu_y} = \frac{E'_y \nu_x}{1 - \nu_x \nu_y}. \tag{6.3c}$$

An elaboration of the description of these constants is presented in Section 6.3.

Since the assumption of isotropy does not enter into the derivation of the expressions for strain, we may use the expressions given by Eqs. (1.7):

$$\epsilon_x = -\mathcal{z}\,\frac{\partial^2 w}{\partial x^2} \tag{6.4a}$$

$$\epsilon_y = -\mathcal{z}\,\frac{\partial^2 w}{\partial y^2} \tag{6.4b}$$

$$\gamma_{xy} = -2\mathcal{z}\,\frac{\partial^2 w}{\partial x\,\partial y}. \tag{6.4c}$$

Substitution of Eqs. (6.4) into (6.1) yields

$$\sigma_x = -\hat{\jmath} \left(E_x \frac{\partial^2 w}{\partial x^2} + E_{xy} \frac{\partial^2 w}{\partial y^2} \right) \tag{6.5a}$$

$$\sigma_y = -\hat{\jmath} \left(E_{xy} \frac{\partial^2 w}{\partial x^2} + E_y \frac{\partial^2 w}{\partial y^2} \right) \tag{6.5b}$$

$$\tau_{xy} = -2G\hat{\jmath} \frac{\partial w}{\partial x \partial y}. \tag{6.5c}$$

The expression for the bending moment stress resultant, M_x, may be derived as follows.

$$M_x = \int_{-h/2}^{h/2} \sigma_x \hat{\jmath} d\hat{\jmath}$$

$$= -\left(E_x \frac{\partial^2 w}{\partial x^2} + E_{xy} \frac{\partial^2 w}{\partial y^2} \right) \int_{-h/2}^{h/2} \hat{\jmath}^2 d\hat{\jmath}$$

$$= -\frac{h^3}{12} \left(E_x \frac{\partial^2 w}{\partial x^2} + E_{xy} \frac{\partial^2 w}{\partial y^2} \right) \tag{6.6a}$$

Similarly, we may derive the expressions for the stress resultants M_y and M_{xy} and obtain

$$M_y = -\frac{h^3}{12} \left(E_y \frac{\partial^2 w}{\partial y^2} + E_{xy} \frac{\partial^2 w}{\partial x^2} \right) \tag{6.6b}$$

and

$$M_{xy} = \frac{2Gh^3}{12} \frac{\partial^2 w}{\partial x \partial y}. \tag{6.6c}$$

We next introduce the following notation to be compatible with the analogous equations from isotropic plate theory.

$$D_x = \frac{h^3 E_x}{12} \tag{6.7a}$$

$$D_y = \frac{h^3 E_y}{12} \tag{6.7b}$$

$$D_{xy} = \frac{h^3 E_{xy}}{12} \tag{6.7c}$$

$$G_{xy} = \frac{h^3 G}{12} \qquad (6.7d)$$

Substitution of Eqs. (6.7) into (6.6) yields

$$M_x = - \left(D_x \frac{\partial^2 w}{\partial x^2} + D_{xy} \frac{\partial^2 w}{\partial y^2} \right) \qquad (6.8a)$$

$$M_y = - \left(D_y \frac{\partial^2 w}{\partial y^2} + D_{xy} \frac{\partial^2 w}{\partial x^2} \right) \qquad (6.8b)$$

$$M_{xy} = 2G_{xy} \frac{\partial^2 w}{\partial x\,\partial y}. \qquad (6.8c)$$

The assumption of isotropy does not enter into the principle of equilibrium as used in Section 1.7 for the derivation of Eq. (1.13). Thus, Eq. (1.13) applies to orthotropic plates as well as to isotropic plates; therefore, the equation of equilibrium for an orthotropic plate is determined by substituting Eqs. (6.8) into (1.13) to obtain

$$D_x \frac{\partial^4 w}{\partial x^4} + (2D_{xy} + 4G_{xy}) \frac{\partial^4 w}{\partial x^2 \partial y^2} + D_y \frac{\partial^4 w}{\partial y^4} = p$$

or

$$D_x \frac{\partial^4 w}{\partial x^4} + 2H \frac{\partial^4 w}{\partial x^2 \partial y^2} + D_y \frac{\partial^4 w}{\partial y^4} = p \qquad (6.9)$$

where

$$H = D_{xy} + 2G_{xy}. \qquad (6.10)$$

The expressions for the stress resultants Q_{xz} and Q_{yz} given by Eqs. (1.12b) and (1.12c) are also applicable to an orthotropic plate; therefore, we obtain the following expressions for these shear stress resultants by substituting Eqs. (6.8) into (1.12b) and (1.12c).

$$Q_{xz} = - \frac{\partial}{\partial x} \left(D_x \frac{\partial^2 w}{\partial x^2} + H \frac{\partial^2 w}{\partial y^2} \right) \qquad (6.11a)$$

$$Q_{yz} = - \frac{\partial}{\partial y} \left(D_y \frac{\partial^2 w}{\partial y^2} + H \frac{\partial^2 w}{\partial x^2} \right) \qquad (6.11b)$$

6.3 ORTHOTROPIC PROPERTIES

An elaboration on the determination of the elastic constants D_x, D_y, D_{xy}, G_{xy}, and H is appropriate before we attempt to solve the governing differential equation given by Eqs. (6.7). The orthotropic constants E_x', E_y', ν_x, ν_y, and G (G is the same for both orthotropic and isotropic materials) are determined by uniaxial tension tests and shear tests of the material with the same techniques that are used for the determination of elastic constants for isotropic materials. The constants D_x, D_y, D_{xy}, G_{xy}, and H then may be determined from Eqs. (6.3), (6.7), and (6.10) as follows.

$$D_x = \frac{h^3 E_x}{12} = \frac{h^3 E_x'}{12(1 - \nu_x \nu_y)}$$

$$D_y = \frac{h^3 E_y}{12} = \frac{h^3 E_y'}{12(1 - \nu_x \nu_y)}$$

$$D_{xy} = \frac{h^3 E_{xy}}{12} = \frac{h^3 \nu_y E_x'}{12(1 - \nu_x \nu_y)} = \frac{h^3 \nu_x E_y'}{12(1 - \nu_x \nu_y)}$$

$$G_{xy} = \frac{h^3 G}{12}$$

$$H = D_{xy} + 2G_{xy}$$

Accuracy in the determination of the elastic constants is often the most crucial part of the problem.

Corrugated plates and plates with stiffeners are frequently treated mathematically as orthotropic plates. Certainly these stiffened plates have varying rigidities in the directions perpendicular and parallel to the stiffeners. Often these plates are modeled by equivalent orthotropic plates with elastic properties equal to the average properties of the various plate components. As an example, the majority of the bending stiffness about the x axis of the corrugated plate in Fig. 6.1 is obtained from the corrugated rib.

However, when this plate is replaced by an equivalent orthotropic plate, the bending stiffness of the flat sheet portions and the ribs are combined and assumed to be distributed evenly across the orthotropic model. Consequently, the net bending stiffness of the corrugated plate and the orthotropic model are made equal. It should be evident that this procedure could yield poor results in certain local areas and yet give a good description of the overall plate stiffness.

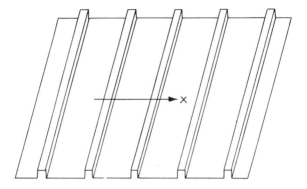

Fig. 6.1   Corrugated plate

The precise meaning of the expression "equivalent orthotropic plate" becomes vague when we model a stiffened plate by an orthotropic plate. We can structure the orthotropic model in such a manner that a certain quantity, such as stress or deflection, in the orthotropic plate matches the corresponding quantity in the stiffened plate. However, we are not guaranteed that stresses will match if deflections are made to match, and vice versa. Of course, the ideal situation is to closely match all the plate variables. Huffington[27] states that we can obtain a satisfactory matching of the overall plate behavior if we can show

$$\frac{s}{a} < < 1$$

$$\frac{s}{b} < < 1$$

in which s is the distance between stiffeners and a and b are the overall plate dimensions. A comprehensive treatment of the techniques for modeling stiffened plates is given by Troitsky.[26] Theoretical approximations of the orthotropic constants often can be determined by various methods; however, these theoretical approximations should be used only when it is not possible to obtain the constants experimentally.

Reinforced Concrete Slabs. The following constants correspond to concrete slabs with steel reinforcement bars in both x and y directions.[1]

$$D_x = \frac{E_c}{1 - \nu_c^2} I_{cx} + \left(\frac{E_s}{E_c} - 1 \right) I_{sx} \qquad (6.12a)$$

$$D_y = \frac{E_c}{1 - \nu_c^2} I_{cy} + \left(\frac{E_s}{E_c} - 1\right) I_{sy} \qquad (6.12b)$$

$$H = \sqrt{D_x D_y} \qquad (6.12c)$$

$$G_{xy} = \frac{1 - \nu_c}{2} \sqrt{D_x D_y} \qquad (6.12d)$$

$$D_{xy} = H - 2G_{xy} = \nu_c \sqrt{D_x D_y} \qquad (6.12c)$$

where

ν_c = Poisson's ratio for concrete

E_c = modulus of elasticity for concrete

I_{cx} and I_{cy} = moments of inertia of the slab material about the neutral axis in a section in which x = constant and y = constant respectively

I_{sx} and I_{sy} = moments of inertia of the reinforcement bars about the neutral axis in a section in which x = constant and y = constant respectively.

Plate Reinforced by Equidistant Stiffeners. For stiffeners in one direction located symmetrically about the middle plane of the plate as shown in Fig. 6.2, the orthotropic elastic constants are approximated by[1]

$$D_x = H = \frac{Eh^3}{12(1 - \nu^2)} \qquad (6.13a)$$

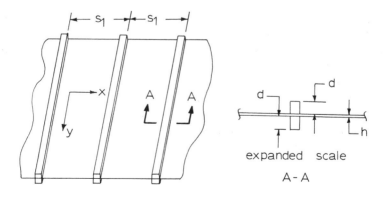

Fig. 6.2. Plate reinforced by equidistant stiffeners

$$D_y = \frac{Eh^3}{12(1 - \nu^2)} + \frac{E' I_1}{S_1} \qquad (6.13b)$$

in which

E and ν = elastic constants of the plating

E' = modulus of elasticity of the stiffeners

I_1 = moment of inertia of the stiffener cross section with respect to the middle surface of the plating

S_1 = spacing between the centerlines of the stiffeners.

If the plate is reinforced by two perpendicular sets of equidistant stiffeners, again assumed to be symmetric with respect to the middle surface of the plating, the orthotropic elastic constants are approximated by[1]

$$D_x = \frac{Eh^3}{12(1 - \nu^2)} \times \frac{E' I_2}{S_2} \qquad (6.14a)$$

$$D_y = \frac{Eh^3}{12(1 - \nu^2)} \times \frac{E' I_1}{S_1} \qquad (6.14b)$$

$$H = \frac{Eh^3}{12(1 - \nu^2)} \qquad (6.14c)$$

in which I_2 and S_2 are the moment of inertia and centerline spacing of the stiffeners that run parallel to the x axis. As before, I_1 and S_1 correspond to the stiffeners that run parallel to the y axis.

Plate Reinforced by a Set of Equidistant Ribs. A plate with reinforcing ribs is illustrated in Fig. 6.3.

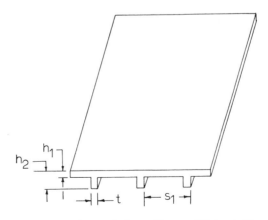

Fig. 6.3. Plate reinforced by equidistant ribs

The orthotropic elastic constants are approximated by[1]

$$D_x = \frac{ES_1 h_1^3}{12\left[S_1 - t + t\left(\dfrac{h_1}{h_2}\right)^3\right]} \tag{6.15a}$$

$$D_y = \frac{EI}{S_1} \tag{6.15b}$$

$$H = 2G'_{xy} + \frac{C}{S_1} \tag{6.15c}$$

$$D_{xy} = 0$$

$$G_{xy} = \frac{H}{2} - \frac{D_{xy}}{2} = \frac{H}{2} \tag{6.15e}$$

in which

C = torsional rigidity of one rib
I = moment of inertia about the neutral axis of a T section as shown by the shaded area in Fig. 6.3.

Corrugated Plates. The determination of the orthotropic constants corresponding to corrugated plates is particularly involved because each new corrugated shape must be treated individually.

The elastic constants for the very special corrugated sheet in Fig. 6.4 are[1]

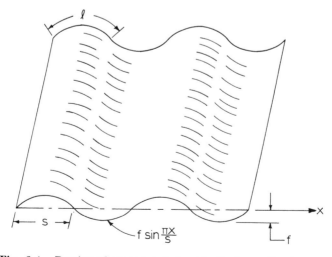

Fig. 6.4. Portion of corrugated sheet in the form of a sine wave

$$D_x = \frac{S}{\ell} \frac{Eh^3}{12(1 - \nu^2)} \qquad (6.16a)$$

$$D_y = EI \qquad (6.16b)$$

$$H = \frac{\ell}{S} \frac{Eh^3}{12(1 - \nu)} \qquad (6.16c)$$

$$D_{xy} = D \qquad (6.16d)$$

$$G_{xy} = \frac{H}{2} - \frac{D_{xy}}{2} = \frac{H}{2} \qquad (6.16e)$$

in which

$$\ell = S \left(1 + \frac{\pi^2 f^2}{4S^2} \right)$$

$$I = \frac{f^2 h}{2} \left[1 - \frac{.81}{1 + 2.5 \left(\frac{f}{2S} \right)^2} \right].$$

McFarland[28] has presented rather lengthy techniques whereby the orthotropic elastic constants can be approximated for any corrugated plate. Again it is recommended that these constants be determined by experimentation whenever possible.

6.4 THE NAVIER SOLUTION FOR ORTHOTROPIC PLATES

Let's consider a simply supported orthotropic plate with the arbitrary transverse loading $p(x,y)$. After we have determined the orthotropic constants as described in Section 6.3, we can use a modified Navier solution to determine the lateral deflections. We assume the solution to be of the form

$$w = \sum_{m=1}^{\infty} \sum_{n=1}^{\infty} w_{mn} \sin \frac{m\pi x}{a} \sin \frac{n\pi y}{b} \qquad (6.17)$$

which satisfies the natural boundary conditions of zero moment and the forced boundary conditions of zero deflection along all four edges. If the transverse load is such that it can be expanded into a Fourier series, we have

$$p(x,y) = \sum_{m=1}^{\infty} \sum_{n=1}^{\infty} p_{mn} \sin \frac{m\pi x}{a} \sin \frac{n\pi y}{b} \qquad (6.18)$$

in which

$$p_{mn} = \frac{4}{ab} \int_0^b \int_0^a p(x,y) \sin \frac{m\pi x}{a} \sin \frac{n\pi y}{b} \, dxdy. \qquad (6.19)$$

Substitution of Eq. (6.18) into (6.9) gives

$$D_x \frac{\partial^4 w}{\partial x^4} + 2H \frac{\partial^4 w}{\partial x^2 \partial y^2} + D_y \frac{\partial^4 w}{\partial y^4}$$

$$= \sum_{m=1}^{\infty} \sum_{n=1}^{\infty} p_{mn} \sin \frac{m\pi x}{a} \sin \frac{n\pi y}{b}. \qquad (6.20)$$

If Eq. (6.17) is to define the lateral deflections, it must satisfy Eq. (6.20); therefore, we must have

$$\sum_{m=1}^{\infty} \sum_{n=1}^{\infty} \left[w_{mn} \left(\frac{m^4 \pi^4}{a^4} D_x + 2H \frac{m^2 n^2 \pi^4}{a^2 b^2} + \frac{n^4 \pi^4}{b^4} D_y \right) - p_{mn} \right]$$

$$\times \sin \frac{m\pi x}{a} \sin \frac{n\pi y}{b} = 0. \qquad (6.21)$$

Equation (6.21) is satisfied for all values of x and y if

$$w_{mn} \left(\frac{m^4 \pi^4}{a^4} D_x + 2H \frac{m^2 n^2 \pi^4}{a^2 b^2} + \frac{n^4 \pi^4}{b^4} \right) - p_{mn} = 0$$

or

$$w_{mn} = \frac{p_{mn}}{\dfrac{m^4 \pi^4}{a^4} D_x + 2H \dfrac{m^2 n^2 \pi^4}{a^2 b^2} + \dfrac{n^4 \pi^4}{b^4} D_y}. \qquad (6.22)$$

Thus, the solution becomes

$$w = \sum_{m=1}^{\infty} \sum_{n=1}^{\infty} \left(\frac{p_{mn}}{\dfrac{m^4 \pi^4}{a^4} D_x + 2H \dfrac{m^2 n^2 \pi^4}{a^2 b^2} + \dfrac{n^4 \pi^4}{b^4} D_y} \right) \sin \frac{m\pi x}{a} \sin \frac{n\pi y}{b}$$

or

$$w = \sum_{m=1}^{\infty} \sum_{n=1}^{\infty} \left[\frac{\dfrac{4}{ab} \displaystyle\int_0^b \int_0^a p(x,y) \sin \dfrac{m\pi x}{a} \sin \dfrac{n\pi y}{b} \, dxdy}{\dfrac{m^4 \pi^4}{a^4} D_x + 2H \dfrac{m^2 n^2 \pi^4}{a^2 b^2} + \dfrac{n^4 \pi^4}{b^4} D_y} \right]$$

$$\times \sin \frac{m\pi x}{a} \sin \frac{n\pi y}{b} . \qquad (6.23)$$

If the material is isotropic, we compare Eqs. (1.9) with (6.1) and conclude

$$E_x = E_y = \frac{E}{1 - \nu^2} \qquad (6.24a)$$

$$E_{xy} = \frac{\nu E}{1 - \nu^2} . \qquad (6.24b)$$

According to Eqs. (6.7) and (6.10) we have

$$D_x = D_y = \frac{Eh^3}{12(1 - \nu^2)} = D \qquad (6.25a)$$

$$D_{xy} = \frac{E\nu h^3}{12(1 - \nu^2)} = \nu D \qquad (6.25b)$$

$$G_{xy} = \frac{Gh^3}{12} = \frac{h^3 E}{24(1 + \nu)} \qquad (6.25c)$$

$$H = \frac{E\nu h^3}{12(1 - \nu^2)} + \frac{h^3 E}{12(1 + \nu)} = D. \qquad (6.25d)$$

Substitution of Eqs. (6.25a) and (6.25b) into Eq. (6.23) gives

$$w = \sum_{m=1}^{\infty} \sum_{n=1}^{\infty} \frac{p_{mn}}{\pi^4 D \left[\left(\dfrac{m}{a} \right)^2 + \left(\dfrac{n}{b} \right)^2 \right]^2} \sin \frac{m\pi x}{a} \sin \frac{n\pi y}{b} \qquad (6.26)$$

which is precisely the equation for lateral deflections that we determined for an isotropic plate.

In this section we have considered orthotropic plates with simply supported edges. It should be pointed out that the assumption of simply supported edges is not as limiting as it might seem. In actual practice engineers often simplify the model of the plate they are analyzing to a simply supported plate. It is reasonably correct to assume that many plates are treated as simply supported plates. There are several reasons why this is a reasonable assumption. First, if the plate is limited by maximum deflections, the analysis of the simply supported plate will indicate more deflection than is actually encountered by a partially constrained plate. Hence, a safety factor is built into the solution. Second, an edge which is restrained must rotate only a very small amount to behave as a hinged plate. The flexibility of the support structure, oversize fastener holes, and looseness of the fasteners are among the conditions which may result in the partially restrained plate exhibiting lateral deflections very nearly the same as a hinged plate. We also sometimes find that an analysis which yields results that are of the order of magnitude of the actual lateral deflections is sufficient. Thus, the relative ease in which the hinged plate can be analyzed compared to the plate with restrained edges often makes the analysis of the hinged plate desirable.

6.5 THE LEVY SOLUTION FOR ORTHOTROPIC PLATES

Again, we can solve the problem of a plate simply supported on two opposite edges and arbitrarily supported on the other two opposite sides with the method proposed by Levy (see Section 3.2). We assume the solution to consist of two parts:

$$w = w_h + w_p \qquad (6.27)$$

in which w_h is the homogeneous part and w_p is the particular part of the solution to Eq. (6.9).

As an example, let us consider an arbitrarily loaded orthotropic plate that is simply supported along the edges $x = 0$ and $x = a$ as shown in Fig. 6.5. The boundary conditions along the edges $y = 0$ and $y = b$ are arbitrary.

The homogeneous form of Eq. (6.9) is

$$D_x \frac{\partial^4 w_h}{\partial x^4} + 2H \frac{\partial^4 w_h}{\partial x^2 \partial y^2} + D_y \frac{\partial^4 w_h}{\partial y^4} = 0. \qquad (6.28)$$

Consistent with the technique prescribed by Levy, we select the form of the homogeneous solution to be

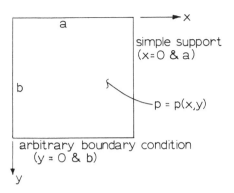

Fig. 6.5. Orthotropic plate with two opposite sides simply supported

$$w_h = \sum_{m=1}^{\infty} f_m(y) \sin \frac{m\pi x}{a} \qquad (6.29)$$

which satisfies the boundary conditions along the simply supported edges. Substitution of Eq. (6.29) into (6.28) yields

$$\sum_{m=1}^{\infty} \left[D_x \left(\frac{m\pi}{a} \right)^4 f_m - 2H \left(\frac{m\pi}{a} \right)^2 \frac{d^2 f_m}{dy^2} + D_y \frac{d^4 f_m}{dy^4} \right] \sin \frac{m\pi x}{a} = 0. \quad (6.30)$$

If Eq. (6.30) is to be satisfied for all values of x, we must have

$$D_x \left(\frac{m\pi}{a} \right)^4 f_m - 2H \left(\frac{m\pi}{a} \right)^2 \frac{d^2 f_m}{dy^2} + D_y \frac{d^4 f_m}{dy^4} = 0. \qquad (6.31)$$

Equation (6.31) is an ordinary differential equation with constant coefficients which we know must have a solution of the form

$$f_m = Ce^{\lambda y} \qquad (6.32)$$

for which C and λ are constants. Substitution of Eq. (6.32) into (6.31) leads to the characteristic equation

$$D_x \left(\frac{m\pi}{a} \right)^4 - 2H\lambda^2 \left(\frac{m\pi}{a} \right)^2 + D_y \lambda^4 = 0. \qquad (6.33)$$

The roots of Eq. (6.33) are

$$\lambda_1 = \frac{m\pi}{a} \sqrt{\frac{1}{D_y} (H + \sqrt{H^2 - D_x D_y})}$$

$$\lambda_2 = \frac{m\pi}{a} \sqrt{\frac{1}{D_y} (H - \sqrt{H^2 - D_x D_y})}$$

$$\lambda_3 = -\frac{m\pi}{a} \sqrt{\frac{1}{D_y} (H + \sqrt{H^2 - D_x D_y})}$$

$$\lambda_4 = -\frac{m\pi}{a} \sqrt{\frac{1}{D_y} (H - \sqrt{H^2 - D_x D_y})}.$$

Thus, the general form of the homogeneous solution becomes

$$w_h = \sum_{m=1}^{\infty} (C_1 e^{\lambda_1 y} + C_2 e^{\lambda_2 y} + C_3 e^{\lambda_3 y} + C_4 e^{\lambda_4 y}) \sin \frac{m\pi x}{a}. \qquad (6.34)$$

The problem is completed when the particular solution is determined and the constants C_1, C_2, C_3, and C_4 are determined from the four boundary conditions on the arbitrary sides. As an example let us develop the general form of the total solution to a plate (with two opposite sides simply supported and the other two sides supported arbitrarily) whose load p(x,y) can be treated as a function of x only. First, we express the loading in terms of the sine series:

$$p(x) = \sum_{m=1}^{\infty} p_m \sin \frac{m\pi x}{a} \qquad (6.35)$$

in which

$$p_m = \frac{2}{a} \int_0^a p(x) \sin \frac{m\pi x}{a} \, dx.$$

Now, we assume the particular solution to be of the form

$$w_p = \sum_{m=1}^{\infty} a_m(y) \sin \frac{m\pi x}{a} \qquad (6.36)$$

which satisfies the simply supported boundary conditions. Substitution of Eqs. (6.35) and (6.36) into the differential equation

$$D_x \frac{\partial^4 w_p}{\partial x^4} + 2H \frac{\partial^4 w_p}{\partial x^2 \partial y^2} + D_y \frac{\partial^4 w_p}{\partial y^4} = p(x,y) \qquad (6.37)$$

yields

$$\sum_{m=1}^{\infty} \left[D_x \left(\frac{m\pi}{a} \right)^4 a_m - 2H \left(\frac{m\pi}{a} \right)^2 \frac{d^2 a_m}{dy^2} + D_y \frac{d^4 a_m}{dy^4} - p_m \right]$$
$$\times \sin \frac{m\pi x}{a} = 0. \qquad (6.38)$$

Equation (6.38) is satisfied for all values of x if

$$D_x \left(\frac{m\pi}{a} \right)^4 a_m - 2H \left(\frac{m\pi}{a} \right)^2 \frac{d^2 a_m}{dy^2} + D_y \frac{d^4 a_m}{dy^4} = p_m. \qquad (6.39)$$

The particular solution of Eq. (6.39) is

$$a_m = \frac{p_m}{D_x} \left(\frac{a}{m\pi} \right)^4. \qquad (6.40)$$

Thus, the expression for w_p becomes

$$w_p = \sum_{m=1}^{\infty} \frac{p_m}{D_x} \left(\frac{a}{m\pi} \right)^4 \sin \frac{m\pi x}{a}. \qquad (6.41)$$

The sum of Eq. (6.34), the homogeneous solution, and Eq. (6.39), the particular solution, gives the general form of the total solution:

$$w = \sum_{m=1}^{\infty} \left[C_1 e^{\lambda_1 y} + C_2 e^{\lambda_2 y} + C_3 e^{\lambda_3 y} + C_4 e^{\lambda_4 y} + \frac{p_m}{D_x} \left(\frac{a}{m\pi} \right)^4 \right]$$
$$\times \sin \frac{m\pi x}{a}. \qquad (6.42)$$

The solution is completed by determining the four constants C_1, C_2, C_3, and C_4 from the four boundary conditions on the two arbitrarily restrained edges.

PROBLEMS

27. Determine a set of equivalent orthotropic constants for the corrugated plate in Fig. 6.6. Assume the edges to be simply supported, and calculate the deflection at the center of the plate when a uniform load $p = p_0$ is applied.

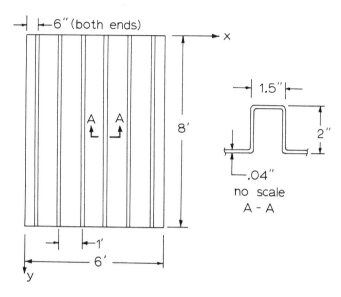

Fig. 6.6

28. Consider the plate and stiffeners in Fig. 6.2 to be constructed of steel. Let's assume

$$s = 10 \text{ in}$$

$$h = .1 \text{ in}$$

$$I = 3 \text{ in}^4$$

$$a = b = 12 \text{ ft.}$$

Determine the maximum allowable uniformly distributed load, $p = p_0$, that can be applied to the plate if $w_{max} = .05$ in.

Chapter 7
ENERGY SOLUTIONS FOR THE LATERAL DEFLECTIONS OF PLATES

7.1 INTRODUCTION

The science of mechanics has developed along two main branches. One branch, which stems from Newton's laws, is referred to as vectorial or Newtonian mechanics. The basic concern of this branch is the recognition of all forces acting on a given body, and the prediction or description of the resulting state of motion or rest. Up to this point this text has taken the Newtonian or vectorial approach. That is, the basic consideration in the development of the governing equations is Newton's law of equilibrium of forces. The second branch, which is based on Bernoulli's principle of virtual work, is referred to as analytical mechanics. In this branch, the vector force is replaced by the work of the force or the potential energy, which are both scalar quantities.

In this chapter we shall present the principle of the first variation of virtual work. We shall then direct this principle to the problem of laterally loaded thin elastic plates and derive expressions for the governing differential equation and the boundary conditions. We shall show that the principle of the first variation of virtual work may be interpreted in the mathematical sense as the principle of stationary work. Stationary values of work are then shown to be realized by the method of Ritz, which leads to approximate solutions for the deflections of plates. A modified method of Ritz is presented, which incorporates Lagrange multipliers for enforcing unusual boundary conditions. Example problems are presented in which deflections are determined by the method of Ritz and by the modified method of Ritz.

7.2 PRINCIPLE OF THE FIRST VARIATION OF VIRTUAL WORK

We may consider a body to consist of a system of N particles whose configuration is specified by the coordinates r_1, r_2, . . ., r_{3N}, where three coordinates are required to specify the position of each particle. Let the forces F_1, F_2, . . ., F_{3N} represent the component forces acting on the particles in the directions of the corresponding coordinates. Imagine that the system is given perfectly arbitrary infinitesimal displacements δr_1, δr_2, . . ., δr_{3N} of the corresponding coordinates. These displacements, called ~~vertical~~ virtual displacements,[16,17] are assumed to occur without passage of time while the forces remain constant. They are not necessarily the displacements that actually take place. It is sufficient that we can imagine them taking place. The mathematical rules of operation for the δ operator are the same as those for the d operator of ordinary differential calculus. However, the d operator refers to an actual infinitesimal change, whereas the δ operator refers to a virtual infinitesimal change.

The expression

$$\delta W = \sum_{i=1} F_i \, \delta r_i \tag{7.1}$$

is defined as the first variation of the work done by the forces as a result of the virtual displacements. (In many texts this expression is simply referred to as the virtual work.) The second and higher variations,[16,20] which are analogous to second and higher derivatives of ordinary differential calculus, are required for the stability analysis of a system and will not be discussed here.

If the body is in equilibrium, then each particle must be in equilibrium, and the resultant force F_i associated with each particular coordinate must be zero. Thus, the first variation of the virtual work must vanish

$$\delta W = 0. \tag{7.2}$$

We conclude that the necessary and sufficient condition for the static equilibrium of a system is that the first variation of the work done by all the forces in moving through their corresponding virtual displacements must be zero.

A more useful form of the principle described by Eq. (7.2) may be obtained by separating the forces into two major categories: (1) external forces or surface forces, and (2) internal forces. Furthermore, the external forces may be classified as either applied forces or reactions. All of these forces may do work in moving through a virtual displacement; therefore, we may express Eq. (7.2) as

$$\delta(W_{EA} + W_{ER} + W_I) = 0 \qquad (7.3a)$$

where W_{EA} is the work of the external applied forces, W_{ER} the work of the external reactions, and W_I the work of the internal forces. Since the δ operator is a linear operator, Eq. (7.3a) may also be expressed as

$$\delta W_{EA} + \delta W_{ER} + \delta W_I = 0. \qquad (7.3b)$$

If we assume that the external reactions, which are sometimes called the forces of constraint, do no work during the virtual displacements, then Eq. (7.3b) becomes

$$\delta W_{EA} + \delta W_I = 0. \qquad (7.4)$$

If the external reactions do no work, the virtual displacements must be limited to a configuration that is compatible with the constraints or the displacement boundary conditions. As an example, the virtual changes in linear and rotational displacements at the clamped end of a cantilever beam must be zero.

We may now state the principle that the necessary and sufficient condition for the static equilibrium of a system with workless constraint forces is that the first variation of the work done by the externally applied forces and the internal forces in moving through virtual displacements compatible with the constraints must be zero. This principle is referred to as the principle of the first variation of virtual work.

7.3 FIRST VARIATION OF VIRTUAL WORK OF PLATE FORCES

According to the principle of the first variation of virtual work presented in Section 7.2 and given by Eq. (7.4), we must now determine the first variation of virtual work of the internal forces of the plate and the external forces applied to the plate.

To determine the first variation of the virtual work of the internal forces of the plate, we shall resort to the summation or integration process of elementary calculus. That is, we shall determine the first variation of the virtual work of the internal forces on a differential volume element of the plate, then integrate over the entire volume to obtain the corresponding expression for the entire plate.

The stresses at the midpoint of each face of the element, which are shown in Fig. 7.1, are obtained by expanding each into a Taylor's series

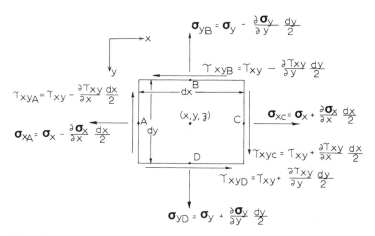

Fig. 7.1. Stresses at the midpoint of each face of a differential plate element

about the point (x,y,γ). The higher-order terms are considered negligible since dx and dy are quantities of infinitesimal magnitude. These stresses and the displacements u and v will generally vary from point to point throughout the plate; therefore, they are functions of x, y, and γ according to Eqs. (1.10) and (1.2).

Let us consider a differential volume element of the plate. The dimensions of the element are dx, dy, and $d\gamma$, and the coordinates of the midpoint of the element are x, y, and γ. The xy midplane of the element is shown in Fig. 7.1. The stresses at the midpoint of the element are defined as σ_x, σ_y, and τ_{xy}.

Since the faces of the differential element are small, the internal forces are obtained by multiplying these stresses by the corresponding areas of the faces on which they act. More exact considerations would lead to higher-order terms that would be considered negligible.

The x and y components of the displacement of the midpoint of the differential element are defined as u and v. The displacement components of the midpoints of the element also are obtained by a Taylor's series expansion about the point (x,y,γ), and are given by the following expressions.

$$u_A = u - \frac{\partial u}{\partial x} \frac{dx}{2}$$

$$v_A = v - \frac{\partial v}{\partial x} \frac{dx}{2}$$

$$u_B = u - \frac{\partial u}{\partial y} \frac{dy}{2}$$

$$v_B = v - \frac{\partial v}{\partial y} \frac{dy}{2}$$

$$u_C = u + \frac{\partial u}{\partial x} \frac{dx}{2}$$

$$v_C = v + \frac{\partial v}{\partial x} \frac{dx}{2}$$

$$u_D = u + \frac{\partial u}{\partial y} \frac{dy}{2}$$

$$v_D = v + \frac{\partial v}{\partial y} \frac{dy}{2}$$

Let us imagine the plate to be modeled with a collection of interconnected ideal elastic elements and ideal mass elements. The ideal elastic element is represented in Fig. 7.1. The first variation of virtual work of the internal plate forces on all the ideal mass elements, which is equal and opposite to the work done by these forces on the ideal elastic elements, can be determined as follows.

$$\delta W_I = -\int \sigma_{xC}\, dy\, d\mathfrak{z}\, \delta u_C + \int \sigma_{xA}\, dy\, d\mathfrak{z}\, \delta u_A - \int \sigma_{yD}\, dx\, d\mathfrak{z}\, \delta v_D$$
$$+ \int \sigma_{yB}\, dx\, d\mathfrak{z}\, \delta v_B - \int \tau_{xyC}\, dy\, d\mathfrak{z}\, \delta v_C + \int \tau_{xyA}\, dy\, d\mathfrak{z}\, \delta v_A$$
$$- \int \tau_{xyD}\, dx\, d\mathfrak{z}\, \delta u_D + \int \tau_{xyB}\, dx\, d\mathfrak{z}\, \delta u_B$$

or

$$\delta W_I = -\int \left(\sigma_x + \frac{\partial \sigma_x}{\partial x} \frac{dx}{2} \right) dy\, d\mathfrak{z}\, \delta \left(u + \frac{\partial u}{\partial x} \frac{dx}{2} \right)$$
$$+ \int \left(\sigma_x - \frac{\partial \sigma_x}{\partial x} \frac{dx}{2} \right) dy\, d\mathfrak{z}\, \delta \left(u - \frac{\partial u}{\partial x} \frac{dx}{2} \right)$$
$$- \int \left(\sigma_y + \frac{\partial \sigma_y}{\partial y} \frac{dy}{2} \right) dx\, d\mathfrak{z}\, \delta \left(v + \frac{\partial v}{\partial y} \frac{dy}{2} \right)$$
$$+ \int \left(\sigma_y - \frac{\partial \sigma_y}{\partial y} \frac{dy}{2} \right) dx\, d\mathfrak{z}\, \delta \left(v - \frac{\partial v}{\partial y} \frac{dy}{2} \right)$$
$$- \int \left(\tau_{xy} + \frac{\partial \tau_{xy}}{\partial x} \frac{dx}{2} \right) dy\, d\mathfrak{z}\, \delta \left(v + \frac{\partial v}{\partial x} \frac{dx}{2} \right)$$

$$+ \int \left(\tau_{xy} - \frac{\partial \tau_{xy}}{\partial x} \frac{dx}{2} \right) dy \, d\beta \, \delta \left(v - \frac{\partial v}{\partial x} \frac{dx}{2} \right)$$

$$- \int \left(\tau_{xy} + \frac{\partial \tau_{xy}}{\partial y} \frac{dy}{2} \right) dx \, d\beta \, \delta \left(u + \frac{\partial u}{\partial y} \frac{dy}{2} \right)$$

$$+ \int \left(\tau_{xy} - \frac{\partial \tau_{xy}}{\partial y} \frac{dy}{2} \right) dx \, d\beta \, \delta \left(u - \frac{\partial u}{\partial y} \frac{dy}{2} \right) \tag{7.5}$$

Let's look at the term

$$\delta u_C = \delta \left(u + \frac{\partial u}{\partial x} \frac{dx}{2} \right).$$

Since the mathematical rules of operation for the δ operator are the same as those for the d operator of ordinary differential calculus, we may write

$$\delta \left(u + \frac{1}{2} \frac{\partial u}{\partial x} dx \right) = \delta u + \frac{1}{2} \delta \left(\frac{\partial u}{\partial x} \right) dx + \frac{1}{2} \frac{\partial u}{\partial x} \delta(dx).$$

Because we are taking the first variation of the displacement and not the plate element length dx, we have

$$\delta \left(u + \frac{1}{2} \frac{\partial u}{\partial x} dx \right) = \delta u + \frac{1}{2} \delta \left(\frac{\partial u}{\partial x} \right) dx.$$

With this operation in mind, Eq. (7.5) becomes

$$\delta W_I = - \int_V \left[\sigma_x \delta \left(\frac{\partial u}{\partial x} \right) + \sigma_y \delta \left(\frac{\partial v}{\partial y} \right) + \tau_{xy} \delta \left(\frac{\partial v}{\partial x} + \frac{\partial u}{\partial y} \right) \right.$$
$$\left. + \left(\frac{\partial \sigma_x}{\partial x} + \frac{\partial \tau_{xy}}{\partial y} \right) \delta u + \left(\frac{\partial \sigma_x}{\partial y} + \frac{\partial \tau_{xy}}{\partial x} \right) \delta v \right] dV \tag{7.6}$$

where V is the volume of the plate, $dV = dx \, dy \, d\beta$. If we substitute Eqs. (1.6), (1.7b), and (1.7c) into Eq. (7.6), we obtain

$$\delta W_I = - \int_V \left(\sigma_x \delta \epsilon_x + \sigma_y \delta \epsilon_y + \tau_{xy} \delta \gamma_{xy} \right) dV$$

$$- \int_V \left[\left(\frac{\partial \sigma_x}{\partial x} + \frac{\partial \tau_{xy}}{\partial y} \right) \delta u + \left(\frac{\partial \sigma_y}{\partial y} + \frac{\partial \tau_{xy}}{\partial x} \right) \delta v \right] dV. \tag{7.7}$$

If we substitute the expressions for stress given by Eqs. (1.9a), (1.9b), and (1.9d) into Eq. (7.7), the first variation of virtual work becomes

$$\delta W_I = - \int_V \left[\frac{E}{1 - \nu^2} \left(\epsilon_x + \nu \epsilon_y \right) \delta \epsilon_x + \frac{E}{1 - \nu^2} \left(\nu \epsilon_x + \epsilon_y \right) \delta \epsilon_y \right.$$

$$
+ \frac{E}{2(1+\nu)} \, \gamma_{xy}\,\delta\gamma_{xy} \bigg] dV - \int_V \bigg[\bigg(\frac{\partial\sigma_x}{\partial x} + \frac{\partial\tau_{xy}}{\partial y} \bigg)\,\delta u
$$
$$
+ \bigg(\frac{\partial\sigma_y}{\partial y} + \frac{\partial\tau_{xy}}{\partial x} \bigg)\,\delta v \bigg] dV
$$

or after rearranging terms

$$
\delta W_I = - \int_V \bigg\{ \frac{E}{1-\nu^2} \left[\epsilon_x\,\delta\epsilon_x + \epsilon_y\,\delta\epsilon_y + \nu(\epsilon_x\,\delta\epsilon_y + \epsilon_y\,\delta\epsilon_x) \right]
$$
$$
+ \frac{E}{2(1+\nu)} \, \gamma_{xy}\,\delta\gamma_{xy} \bigg\} dV - \int_V \bigg[\bigg(\frac{\partial\sigma_x}{\partial x} + \frac{\partial\tau_{xy}}{\partial y} \bigg)\,\delta u
$$
$$
+ \bigg(\frac{\partial\sigma_y}{\partial y} + \frac{\partial\tau_{xy}}{\partial x} \bigg)\,\delta v \bigg] dV.
$$

Since the mathematical rules of operation for the δ operator are the same as those for the d operator of ordinary differential calculus, we may write $1/2\,\delta\epsilon_x^2 = \epsilon_x\,\delta\epsilon_x$, etc. Thus, the expression for the first variation of virtual work is

$$
\delta W_I = - \int_V \bigg[\frac{E}{1-\nu^2}\,\delta\bigg(\frac{\epsilon_x^2}{2} + \frac{\epsilon_y^2}{2} + \nu\epsilon_x\,\epsilon_y \bigg) + \frac{E}{2(1+\nu)}\,\delta\frac{\gamma_{xy}^2}{2} \bigg] dV
$$
$$
- \int_V \bigg[\bigg(\frac{\partial\sigma_x}{\partial x} + \frac{\partial\tau_{xy}}{\partial y} \bigg)\,\delta u + \bigg(\frac{\partial\sigma_y}{\partial y} + \frac{\partial\tau_{xy}}{\partial x} \bigg)\,\delta v \bigg] dV
$$

or

$$
\delta W_I = -\delta \int_V \frac{1}{2} \bigg[\frac{E}{1-\nu^2}\,(\epsilon_x + \nu\epsilon_y)\,\epsilon_x + \frac{E}{1-\nu^2}\,(\epsilon_y + \nu\epsilon_x)\,\epsilon_y
$$
$$
+ \frac{E}{2(1+\nu)} \, \gamma_{xy}\gamma_{xy} \bigg] dV - \int_V \bigg[\bigg(\frac{\partial\sigma_x}{\partial x} + \frac{\partial\tau_{xy}}{\partial y} \bigg)\,\delta u
$$
$$
+ \bigg(\frac{\partial\sigma_y}{\partial y} + \frac{\partial\tau_{xy}}{\partial x} \bigg)\,\delta v \bigg] dV.
$$

If we once again refer to the expressions for stress given by Eqs. (1.9a), (1.9b), and (1.9d), we obtain

$$
\delta W_I = -\delta \int_V \frac{1}{2}\,(\sigma_x\epsilon_x + \sigma_y\epsilon_y + \tau_{xy}\gamma_{xy})\,dV
$$
$$
- \int_V \bigg[\bigg(\frac{\partial\sigma_x}{\partial x} + \frac{\partial\tau_{xy}}{\partial y} \bigg)\,\delta u + \bigg(\frac{\partial\sigma_y}{\partial y} + \frac{\partial\tau_{xy}}{\partial x} \bigg)\,\delta v \bigg] dV. \tag{7.8}
$$

The conditions of equilibrium are obtained from Fig. 7.1 by requiring that the resultant force on the differential element vanish. If we consider the x and y components of the resultant force separately, we have

$$\sum F_x = 0 = \left(\sigma_x + \frac{\partial \sigma_x}{\partial x}\frac{dx}{2}\right) dy \, dz - \left(\sigma_x - \frac{\partial \sigma_x}{\partial x}\frac{dx}{2}\right) dy \, dz$$
$$+ \left(\tau_{xy} + \frac{\partial \tau_{xy}}{\partial y}\frac{dy}{2}\right) dx \, dz - \left(\tau_{xy} - \frac{\partial \tau_{xy}}{\partial y}\frac{dy}{2}\right) dx \, dz \qquad (7.9a)$$

$$\sum F_y = 0 = \left(\sigma_y + \frac{\partial \sigma_y}{\partial y}\frac{dy}{2}\right) dx \, dz - \left(\sigma_y - \frac{\partial \sigma_y}{\partial y}\frac{dy}{2}\right) dx \, dz$$
$$+ \left(\tau_{xy} + \frac{\partial \tau_{xy}}{\partial x}\frac{dx}{2}\right) dy \, dz - \left(\tau_{xy} - \frac{\partial \tau_{xy}}{\partial x}\frac{dx}{2}\right) dy \, dz. \qquad (7.9b)$$

Equations (7.9) reduce to

$$\frac{\partial \sigma_x}{\partial x} + \frac{\partial \tau_{xy}}{\partial y} = 0 \qquad (7.10a)$$

$$\frac{\partial \sigma_y}{\partial y} + \frac{\partial \tau_{xy}}{\partial x} = 0. \qquad (7.10b)$$

Consequently, the condition of equilibrium, given by Eqs. (7.10), requires the second integral of Eq. (7.8) to vanish. Thus, the first variation of the virtual work of the internal plate forces becomes

$$\delta W_I = -\delta \int_V \frac{1}{2} \left(\sigma_x \epsilon_x + \sigma_y \epsilon_y + \tau_{xy} \gamma_{xy}\right) dV. \qquad (7.11)$$

The integral of Eq. (7.11) is defined as the strain energy, U, of the plate.

$$U = \int_V \frac{1}{2} \left(\sigma_x \epsilon_x + \sigma_y \epsilon_y + \tau_{xy} \gamma_{xy}\right) dV$$

This integral will be of much more use to us in the remainder of this chapter if we express it in terms of the deflection of the middle surface of the plate, w.

If we substitute Eqs. (1.7) and (1.10) into the expression for the strain energy and integrate over z, the strain energy becomes

$$U = \frac{D}{2} \int_A \left\{ \left(\frac{\partial^2 w}{\partial x^2} + \frac{\partial^2 w}{\partial y^2}\right)^2 \right.$$
$$\left. - 2(1 - \nu) \left[\frac{\partial^2 w}{\partial x^2}\frac{\partial^2 w}{\partial y^2} - \left(\frac{\partial^2 w}{\partial x \, \partial y}\right)^2\right] \right\} dx \, dy \qquad (7.12)$$

where A = area of the plate.

If we use the transformations given by Eqs. (2.3), the strain energy for a circular plate is

$$U = \frac{D}{2} \int_A \left[\left(\frac{\partial^2 w}{\partial r^2} + \frac{1}{r} \frac{\partial w}{\partial r} + \frac{1}{r^2} \frac{\partial^2 w}{\partial \theta^2} \right)^2 - 2(1 - \nu) \frac{\partial^2 w}{\partial r^2} \right.$$
$$\left. \times \left(\frac{1}{r} \frac{\partial w}{\partial r} + \frac{1}{r^2} \frac{\partial^2 w}{\partial \theta^2} \right) + 2(1 - \nu) \left(\frac{1}{r} \frac{\partial^2 w}{\partial r \partial \theta} - \frac{1}{r^2} \frac{\partial w}{\partial \theta} \right)^2 \right] r \, dr \, d\theta. \quad (7.13)$$

The first variation of virtual work of an externally applied distributed transverse load, p(x,y), is

$$\delta W_{EA} = \int_{A_1} p(x,y) \, dx \, dy \, \delta w \quad (7.14)$$

where A_1 = the surface area of the plate over which the load is distributed. Since the forces are assumed to remain constant during a virtual displacement, we may place the δ operator before the integral sign, and Eq. (7.14) becomes

$$\delta W_{EA} = \delta \int_{A_1} p(x,y) \, w(x,y) \, dx \, dy. \quad (7.15)$$

The expression for the first variation of virtual work of all the plate forces is obtained by combining Eq. (7.15) with (7.12) as follows.

$$\delta W = \delta \int_{A_1} p(x,y) \, w(x,y) \, dx \, dy - \frac{D}{2} \delta \int_A \left\{ \left(\frac{\partial^2 w}{\partial x^2} + \frac{\partial^2 w}{\partial y^2} \right)^2 \right.$$
$$\left. - 2(1 - \nu) \left[\frac{\partial^2 w}{\partial x^2} \frac{\partial^2 w}{\partial y^2} - \left(\frac{\partial^2 w}{\partial x \partial y} \right)^2 \right] \right\} dx \, dy \quad (7.16)$$

Any loading

If the external load is a concentrated load P at $x = x_0$ and $y = y_0$, the first variation of the virtual work is

$$\delta W_{EA} = P \delta w(x_0, y_0) \quad (7.17)$$

and the expression for the first variation of the virtual work of all the plate forces is

$$\delta W = P \delta w(x_0, y_0) - \frac{D}{2} \delta \int_A \left\{ \left(\frac{\partial^2 w}{\partial x^2} + \frac{\partial^2 w}{\partial y^2} \right)^2 \right.$$
$$\left. - 2(1 - \nu) \left[\frac{\partial^2 w}{\partial x^2} \frac{\partial^2 w}{\partial y^2} - \left(\frac{\partial^2 w}{\partial x \partial y} \right)^2 \right] \right\} dx \, dy. \quad (7.18)$$

Concentrated loading

7.4 GOVERNING EQUATION AND BOUNDARY CONDITIONS FROM PRINCIPLE OF FIRST VARIATION OF VIRTUAL WORK

According to the principle of the first variation of virtual work described in Section 7.2, the first variation of work done by the forces applied externally to the plate and the internal plate forces in moving through virtual displacements compatible with the constraints must vanish at a position of equilibrium. Thus, Eq. (7.16) is set equal to zero.

$$\delta \int_{A_1} \text{pwdx}\,dy - \frac{D}{2} \delta \int_A \left[\left(\frac{\partial^2 w}{\partial x^2}\right)^2 + 2 \frac{\partial^2 w}{\partial x^2} \frac{\partial^2 w}{\partial y^2} + \left(\frac{\partial^2 w}{\partial y^2}\right)^2 \right] dx\,dy$$
$$+ D(1 - \nu)\delta \int_A \left[\frac{\partial^2 w}{\partial x^2} \frac{\partial^2 w}{\partial y^2} - \left(\frac{\partial^2 w}{\partial x\,\partial y}\right)^2 \right] dx\,dy = 0 \quad (7.19)$$

The third integral of Eq. (7.19) may be expressed in a slightly different form so that we have

$$\delta \int_{A_1} \text{pwdx}\,dy - \frac{D}{2} \delta \int_A \left[\left(\frac{\partial^2 w}{\partial x^2}\right)^2 + 2 \frac{\partial^2 w}{\partial x^2} \frac{\partial^2 w}{\partial y^2} + \left(\frac{\partial^2 w}{\partial y^2}\right)^2 \right] dx\,dy$$
$$+ D(1 - \nu)\delta \int_A \left[\frac{\partial}{\partial x}\left(\frac{\partial w}{\partial x} \frac{\partial^2 w}{\partial y^2}\right) - \frac{\partial}{\partial y}\left(\frac{\partial w}{\partial x} \frac{\partial^2 w}{\partial x\,\partial y}\right) \right] dx\,dy = 0. \quad (7.20)$$

Before we proceed, let us recall from calculus Green's theorem in a plane,[19] which is expressed as follows.

$$\oint_C [F(x,y)\,dx + G(x,y)\,dy] = \int_R \left[\frac{\partial G(x,y)}{\partial x} - \frac{\partial F(x,y)}{\partial y} \right] dx\,dy \quad (7.21)$$

where R is a closed region in the x-y plane, bounded by the curve C such that the region is on the right if an advancement is made along the curve in the positive direction, Fig. 7.2, and F and G are continuous functions with continuous first partial derivatives.

If we apply Green's theorem given by Eq. (7.21) to the third integral of Eq. (7.20), we obtain

$$\delta W = \delta \int_{A_1} \text{pwdx}\,dy - \frac{D}{2} \delta \int_A \left[\left(\frac{\partial^2 w}{\partial x^2}\right)^2 + 2 \frac{\partial^2 w}{\partial x^2} \frac{\partial^2 w}{\partial y^2} \right.$$
$$\left. + \left(\frac{\partial^2 w}{\partial y^2}\right)^2 \right] dx\,dy + D(1 - \nu)\delta \oint_C \frac{\partial w}{\partial x} \frac{\partial^2 w}{\partial x\,\partial y}\,dx$$
$$+ D(1 - \nu)\delta \oint_C \frac{\partial w}{\partial x} \frac{\partial^2 w}{\partial y^2}\,dy = 0. \quad (7.22)$$

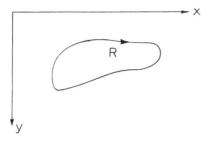

Fig. 7.2

The reason for expressing the third integral of Eq. (7.20) as a line integral will be explained in the last paragraph of this section.

If we investigate Eq. (7.22) we see that the work function W may be expressed as a function of w(x,y) and several partial derivatives of w(x,y)

$$W = W\left(w, \frac{\partial w}{\partial x}, \frac{\partial^2 w}{\partial x^2}, \frac{\partial^2 w}{\partial y^2}, \frac{\partial^2 w}{\partial x\, \partial y}\right).$$

The function W(x,y) may receive two kinds of increments: (1) the differential increment dW caused by the increments dx and dy, and (2) the first variation[20] δW which represents an imaginary increment added to the function W without a change in x and y. Since the rules of operation for the δ operator are the same as those for the d operator of ordinary differential calculus, the first variation of W is defined as

$$\delta W = \frac{\partial W}{\partial x} \delta w + \frac{\partial W}{\partial \left(\dfrac{\partial w}{\partial x}\right)} \delta\left(\frac{\partial w}{\partial x}\right) + \frac{\partial W}{\partial \left(\dfrac{\partial^2 w}{\partial x^2}\right)} \delta\left(\frac{\partial^2 w}{\partial x^2}\right)$$

$$+ \frac{\partial W}{\partial \left(\dfrac{\partial^2 w}{\partial y^2}\right)} \delta\left(\frac{\partial^2 w}{\partial y^2}\right) + \frac{\partial W}{\partial \left(\dfrac{\partial^2 w}{\partial x\, \partial y}\right)} \delta\left(\frac{\partial^2 w}{\partial x\, \partial y}\right). \qquad (7.23)$$

The first variation δw(x,y) represents an imaginary infinitesimal increment added to the function w(x,y) without a change in x and y. The first variation $\delta[\partial w(x,y)/\partial x]$ represents an imaginary infinitesimal increment added to the function $\partial w(x,y)/\partial x$ without a change in x and y. The other first variations in Eq. (7.23) may be described similarly.

According to the definition given by Eq. (7.23), the expression for δW from Eq. (7.22) may now be written as

$$\int_{A_1} p\delta w\,dx\,dy \; - \; D \int_A \left[\frac{\partial^2 w}{\partial x^2} \delta\left(\frac{\partial^2 w}{\partial x^2}\right) + \frac{\partial^2 w}{\partial x^2} \delta\left(\frac{\partial^2 w}{\partial y^2}\right)\right.$$

$$\left. + \frac{\partial^2 w}{\partial y^2} \delta\left(\frac{\partial^2 w}{\partial x^2}\right) + \frac{\partial^2 w}{\partial y^2} \delta\left(\frac{\partial^2 w}{\partial y^2}\right)\right]dx\,dy$$

$$+ D(1 - \nu) \oint_C \left[\frac{\partial w}{\partial x} \delta\left(\frac{\partial^2 w}{\partial x\,\partial y}\right) + \frac{\partial^2 w}{\partial x\,\partial y} \delta\left(\frac{\partial w}{\partial x}\right)\right] dx$$

$$+ D(1 - \nu) \oint_C \left[\frac{\partial w}{\partial x} \delta\left(\frac{\partial^2 w}{\partial y^2}\right) + \frac{\partial^2 w}{\partial y^2} \delta\left(\frac{\partial w}{\partial x}\right)\right] dy = 0. \quad (7.24)$$

We must now integrate every term of Eq. (7.24), except the first, by parts. In this process, we must remember that the δ operator and the $\partial/\partial x$ and $\partial/\partial y$ operators are interchangeable since $w(w,y)$ is assumed to be a continuous function for which the derivatives exist. For example:

$$\delta \frac{\partial w}{\partial x} = \frac{\partial}{\partial x} (\delta w)$$

$$\delta \frac{\partial^2 w}{\partial x^2} + \frac{\partial^2}{\partial x^2} (\delta w)$$

$$\frac{\partial^2 w}{\partial x\,\partial y} = \frac{\partial^2 w}{\partial y\,\partial x}$$

etc.

To illustrate this process of integration by parts, we shall consider the following two examples. Refer to the coordinate system in Fig. 1.1.

The first term of the second integral of Eq. (7.4) is integrated by parts as follows.

$$\int_A \frac{\partial^2 w}{\partial x^2} \delta\left(\frac{\partial^2 w}{\partial x^2}\right) dx\,dy = \int_0^b \left[\frac{\partial^2 w}{\partial x^2} \delta\left(\frac{\partial w}{\partial x}\right)\right]_{x=0}^{x=a}$$

$$- \int_0^a \frac{\partial^3 w}{\partial x^3} \delta\left(\frac{\partial w}{\partial x}\right) dx \Bigg] dy = \int_0^b \frac{\partial^2 w}{\partial x^2} \delta\left(\frac{\partial w}{\partial x}\right)\Bigg|_{x=0}^{x=a} dy$$

$$- \int_0^b \frac{\partial^3 w}{\partial x^3} \delta w \Bigg|_{x=0}^{x=a} dy + \int_0^b \int_0^a \frac{\partial^4 w}{\partial x^4} \delta w\,dx\,dy$$

The first term of the third integral of Eq. (7.24) is integrated by parts as follows.

$$\oint_C \frac{\partial w}{\partial x} \delta\left(\frac{\partial^2 w}{\partial x \partial y}\right) dx = \int_0^a \frac{\partial w}{\partial x} \delta\left(\frac{\partial^2 w}{\partial x \partial y}\right)\Bigg|_{y=0} dx$$

$$- \int_0^a \frac{\partial w}{\partial x} \delta\left(\frac{\partial^2 w}{\partial x \partial y}\right)\Bigg|_{y=b} dx = \frac{\partial w}{\partial x} \delta\left(\frac{\partial w}{\partial y}\right)\Bigg|_{y=0}\Bigg|_{x=0}^{x=a} \not{dx}$$

$$- \int_0^a \frac{\partial^2 w}{\partial x^2} \delta\left(\frac{\partial w}{\partial y}\right)\Bigg|_{y=0} dx - \frac{\partial w}{\partial x} \delta\left(\frac{\partial w}{\partial y}\right)\Bigg|_{y=b}\Bigg|_{x=0}^{x=a}$$

$$+ \int_0^a \frac{\partial^2 w}{\partial x^2} \delta\left(\frac{\partial w}{\partial y}\right)\Bigg|_{y=b} dx = \int_0^a \frac{\partial^2 w}{\partial x^2} \delta\left(\frac{\partial w}{\partial y}\right)\Bigg|_{y=0}^{y=b} dx$$

$$+ \frac{\partial w}{\partial x} \delta\left(\frac{\partial w}{\partial y}\right)\Bigg|_{y=0}\Bigg|_{x=0}^{x=a} - \frac{\partial w}{\partial x} \delta\left(\frac{\partial w}{\partial y}\right)\Bigg|_{y=b}\Bigg|_{x=0}^{x=a}$$

$$= \int_0^a \frac{\partial^2 w}{\partial x^2} \delta\left(\frac{\partial w}{\partial y}\right)\Bigg|_{y=0}^{y=b} dx - \frac{\partial w}{\partial x} \delta\left(\frac{\partial w}{\partial y}\right)_{\substack{x=0 \\ y=0}} + \frac{\partial w}{\partial x} \delta\left(\frac{\partial w}{\partial y}\right)_{\substack{x=a \\ y=0}}$$

$$+ \frac{\partial w}{\partial x} \delta\left(\frac{\partial w}{\partial y}\right)_{\substack{x=0 \\ y=b}} - \frac{\partial w}{\partial x} \delta\left(\frac{\partial w}{\partial y}\right)_{\substack{x=a \\ y=b}}$$

After each term of Eq. (7.24), except the first term, is integrated by parts, we have the following expression for the first variation of virtual work.

$$\delta W = D \int_A \left(\frac{\partial^4 w}{\partial x^4} + 2\frac{\partial^4 w}{\partial x^2 \partial y^2} + \frac{\partial^4 w}{\partial y^4}\right)\overset{-\frac{P}{D}}{\delta w\, dx\, dy} + D \int_0^b \left(\frac{\partial^2 w}{\partial x^2}\right.$$

$$+ \left. \nu \frac{\partial^2 w}{\partial y^2}\right) \delta\left(\frac{\partial w}{\partial x}\right)\Bigg|_{x=0}^{x=a} dy + D \int_0^b \left(\frac{\partial^3 w}{\partial x^3}\right.$$

$$+ \left. (2 - \nu) \frac{\partial^3 w}{\partial x \partial y^2}\right) \delta w\Bigg|_{x=0}^{x=a} dy + D \int_0^a \left(\frac{\partial^2 w}{\partial y^2}\right.$$

$$+ \left. \nu \frac{\partial^2 w}{\partial x^2}\right) \delta\left(\frac{\partial w}{\partial y}\right)\Bigg|_{y=0}^{y=b} dx + D \int_0^a \left(\frac{\partial^3 w}{\partial y^3}\right.$$

$$+ \left. (2 - \nu) \frac{\partial^3 w}{\partial x^2 \partial y}\right) \delta w\Bigg|_{y=0}^{y=b} dx - 2(1 - \nu) D \frac{\partial^2 w}{\partial x \partial y} \delta w\Bigg|_{\substack{x=0 \\ y=0}}$$

$$+ 2(1 - \nu)D\frac{\partial^2 w}{\partial x \, \partial y} \, \delta w \bigg|_{\substack{x = a \\ y = 0}} \qquad - 2(1 - \nu) \, D \, \frac{\partial^2 w}{\partial x \, \partial y} \, \delta w \bigg|_{\substack{x = a \\ y = b}}$$

$$+ 2(1 - \nu) \, D \, \frac{\partial^2 w}{\partial x \, \partial y} \, \delta w \bigg|_{\substack{x = 0 \\ y = b}} \qquad = 0 \qquad (7.25)$$

Since the first variations δw, $\delta(\partial w / \partial x)$, and $\delta(\partial w / \partial y)$ are independent, Eq. (7.5) is satisfied as follows.

7.25

$$\frac{\partial^4 w}{\partial x^4} + 2 \, \frac{\partial^4 w}{\partial x^2 \, \partial y^2} + \frac{\partial^4 w}{\partial y^4} - \frac{p}{D} = 0 \left. \right\} \text{ over the entire area, A} \quad (7.26)$$

$$\text{Either} \quad \frac{\partial^2 w}{\partial x^2} + \nu \, \frac{\partial^2 w}{\partial y^2} = 0 \tag{7.27a}$$

$$\text{or} \quad \delta \left(\frac{\partial w}{\partial x} \right) = 0 \qquad \qquad \text{along the} \tag{7.27b}$$

edges $x = 0$

$$\text{and either} \quad \frac{\partial^3 w}{\partial x^3} + (2 - \nu) \, \frac{\partial^3 w}{\partial x \, \partial y^2} = 0 \qquad \text{and } x = a \tag{7.27c}$$

$$\text{or} \quad \delta w = 0 \tag{7.27d}$$

$$\text{Either} \quad \frac{\partial^2 w}{\partial y^2} + \nu \, \frac{\partial^2 w}{\partial x^2} = 0 \tag{7.28a}$$

$$\text{or} \quad \delta \left(\frac{\partial w}{\partial y} \right) = 0 \qquad \qquad \text{along the} \tag{7.28b}$$

edges $y = 0$

$$\text{and either} \quad \frac{\partial^3 w}{\partial y^3} + (2 - \nu) \, \frac{\partial^3 w}{\partial x^2 \, \partial y} = 0 \qquad \text{and } y = b \tag{7.28c}$$

$$\text{or} \quad \delta w = 0 \tag{7.28d}$$

$$\text{Either} \quad 2D(1 - \nu) \, \frac{\partial^2 w}{\partial x \, \partial y} = 0 \left. \right\} \text{ at each corner} \tag{7.29a}$$

$$\text{or} \quad \delta w = 0 \tag{7.29b}$$

Equation (7.26) is the governing equation. It is identical to Eq. (1.13) derived in Section 1.8.

Equations (7.27) represent the boundary conditions on the edges $x = 0$ and $x = a$. They may be interpreted as follows.

Simply Supported Edge

$$\frac{\partial^2 w}{\partial x^2} + \nu \ \cancel{\frac{\partial^2 w}{\partial y^2}}^{\ 0} = 0 \quad \ldots \text{[refer to Eq. (7.27a)]}$$

$\delta w = 0$, which implies that w is constant and if w is zero initially, it remains zero   . . . [refer to Eq. (7.27d)]

Clamped Edge

$\delta \left(\dfrac{\partial w}{\partial x} \right) = 0$, which implies that $\partial w / \partial x$ is constant and if $\partial w / \partial x$ is zero initially, it remains zero   . . . [refer to Eq. (7.27b)]

$\delta w = 0$, which implies that w is constant and if w is zero initially, it remains zero   . . . [refer to Eq. (7.27d)]

Free Edge

$$\frac{\partial^2 w}{\partial x^2} + \nu \frac{\partial^2 w}{\partial y^2} = 0, \quad \ldots \text{[refer to Eq. (7.27a)]}$$

$$\frac{\partial^3 w}{\partial x^3} + (2 - \nu) \frac{\partial^3 w}{\partial x \, \partial y^2} = 0, \quad \ldots \text{[refer to Eq. (7.27c)]}$$

Equations (7.28) may be interpreted in a similar manner to obtain the boundary conditions on the edges $y = 0$ and $y = b$.

Equations (7.29) represent the conditions at the corners presented in Section (1.10). For a rectangular plate constrained around the edges in some manner such that $w = 0$, there is required a force at each corner whose magnitude is given by the expression Eq. (7.29a).

The boundary conditions of deflection and slope described by Eqs. (7.27b), (7.27d), (7.28b), and (7.28d) are called forced boundary conditions, since they are mathematical expressions of constraint. The boundary conditions of shear and moment described by Eqs. (7.27a), (7.27c), (7.28a), and (7.28c) are called natural boundary conditions.

Now let us discuss the reason for expressing the third integral of Eq. (7.20) as a line integral. Let's look at the second term of the third integral of Eq. (7.19).

$$\delta \int_A \left(\frac{\partial^2 w}{\partial x \, \partial y} \right)^2 dx \, dy$$

or

$$2 \int_A \frac{\partial^2 w}{\partial x \, \partial y} \ \delta \left(\frac{\partial^2 w}{\partial x \, \partial y} \right) dx \, dy$$

We see that the integrand is a mixed partial derivative which may be integrated by parts two different ways. We may integrate with respect to x first, then with respect to y, or we may perform our integrations in the reverse order. To obtain the entire set of boundary conditions, the integration must be accomplished both ways. To avoid the confusion of this procedure, we introduced Green's theorem which reduced the area integrals to line integrals. The line integrals involved integration by parts only once eliminating the process of integrating by parts using two different orders of integration.

7.5 DEFLECTIONS FROM PRINCIPLE OF STATIONARY WORK BY METHOD OF RITZ

Since the mathematical rules of operation for the δ operator are the same as those for the d operator of ordinary differential calculus, Eq. (7.2)

$$\delta W = 0$$

implies that the value of the work function, W, at a position of equilibrium is a stationary[18] value.

We may elaborate on this idea of a stationary value of the work function by applying the chain rule of elementary calculus as follows.

$$\delta W(r_1, r_2, \ldots, r_{3N}) = 0$$

$$\frac{\partial W}{\partial r_1} \delta r_1 + \frac{\partial W}{\partial r_2} \delta r_2 + \ldots + \frac{\partial W}{\partial r_{3N}} \delta r_{3N} = 0 \qquad (7.30)$$

If the coordinates r_i are an independent set (if they are not we may always transform to an independent set), the coefficients of the virtual displacements δr_i must vanish:

$$\frac{\partial W}{\partial r_i} = 0 \qquad i = 1, 2, \ldots, 3N. \qquad (7.31)$$

Equations (7.31) are the necessary and sufficient conditions for a stationary value of W, providing it has continuous first partial derivatives.

According to the method of Ritz,[20,21] the deflection of the plate is taken in the form of a series

$$w = C_1 \psi_1(x,y) + C_2 \psi_2(x,y) + \ldots + C_n \psi_n(x,y) = \sum_{i=1}^{n} C_i \psi_i(x,y). \quad (7.32)$$

If we substitute Eq. (7.32) into the expression for the work W, and the integrations are performed, the work becomes a second degree function of the coefficients C_1, C_2, . . ., C_n. To satisfy the conditions for a stationary value of W

$$\frac{\partial W}{\partial C_1} = 0 \qquad \frac{\partial W}{\partial C_2} = 0 \qquad \cdots \qquad \frac{\partial W}{\partial C_n} = 0 \qquad (7.33)$$

which results in a system of n linear equations in the unknown coefficients C_1, C_2, . . ., C_n that can easily be solved.

The functions ψ_1, ψ_2, . . ., ψ_n are chosen such that they are compatible with the constraints in accordance with the theorem of the first variation of virtual work. That is, they must satisfy the forced boundary conditions (the prescribed conditions of deflection and slope). If the set of functions is complete,[3,4] the solution converges in the mean-square sense[18] to the exact solution by taking n infinitely large, and the natural boundary conditions are automatically satisfied. Otherwise, the solution is approximate. The set of functions is only required to satisfy the forced boundary conditions; however, if the natural boundary conditions are also satisfied, convergence is more rapid.

Example

Determine the deflection of a simply supported rectangular plate subjected to a uniform lateral load, $p(x,y) = p_0$, as shown in Fig. 7.3.

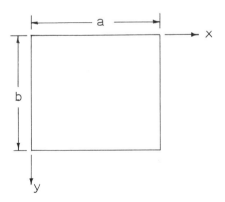

Fig. 7.3. Simply supported plate with uniform lateral load, $p(x,y) = p_0$

The work function, obtained from Eq. (7.22), is

$$W = \int_{A_1} p_0 w \, dx \, dy - \frac{D}{2} \int_A \left[\left(\frac{\partial^2 w}{\partial x^2} \right)^2 + 2 \frac{\partial^2 w}{\partial x^2} \frac{\partial^2 w}{\partial y^2} + \left(\frac{\partial^2 w}{\partial y^2} \right)^2 \right] dx \, dy$$

$$A \sin \frac{n\pi x}{a} \qquad \sum_{n=odd}^{\infty} A_{mn} \left(1 - \cos \frac{n\pi x}{2a} \right) \qquad A x (x-a)^2 \qquad A x^2 (x-a)^2$$

$$+ D(1 - \nu) \oint_C \frac{\partial w}{\partial x} \frac{\partial^2 w}{\partial x \partial y} \, dx + D(1 - \nu) \oint_C \frac{\partial w}{\partial x} \frac{\partial^2 w}{\partial y^2} \, dy.$$

For a simply supported plate, we see that the last two integrals must vanish; thus, the work function reduces to the following expression.

$$W = \int_{A_1} p_0 w \, dx \, dy - \frac{D}{2} \int_A \left[\left(\frac{\partial^2 w}{\partial x^2} \right)^2 + 2 \frac{\partial^2 w}{\partial x^2} \frac{\partial^2 w}{\partial y^2} + \left(\frac{\partial^2 w}{\partial y^2} \right)^2 \right] dx \, dy$$

The deflection of the plate is taken in the form of a double trigonometric sine series as follows.

$$w = \sum_{m=1}^{\infty} \sum_{n=1}^{\infty} A_{mn} \sin \frac{m \pi x}{a} \sin \frac{n \pi y}{b}$$

This form for the deflection satisfies the forced boundary conditions of zero deflection (and non-zero slope) on all edges and will converge to the exact solution by taking n infinitely large. This form also has the advantage of being a series of orthogonal functions which greatly simplifies the integration of the work function.

Upon substitution of the expression for the deflection into the work function, we have

$$W = p_0 \int_0^b \int_0^a \sum_{m=1}^{\infty} \sum_{n=1}^{\infty} A_{mn} \sin \frac{m \pi x}{a} \sin \frac{n \pi y}{b} \, dx \, dy$$

$$- \frac{D}{2} \int_0^b \int_0^a \left[\sum_{m=1}^{\infty} \sum_{n=1}^{\infty} A_{mn} \left(\frac{m \pi}{a} \right)^2 \sin \frac{m \pi x}{a} \sin \frac{n \pi y}{b} \right]^2 dx \, dy$$

$$- D \int_0^b \int_0^a \left[\sum_{m=1}^{\infty} \sum_{n=1}^{\infty} A_{mn} \left(\frac{m \pi}{a} \right)^2 \sin \frac{m \pi x}{a} \sin \frac{n \pi y}{b} \right]$$

$$\times \left[\sum_{m=1}^{\infty} \sum_{n=1}^{\infty} A_{mn} \left(\frac{n \pi}{b} \right)^2 \sin \frac{m \pi x}{a} \sin \frac{n \pi y}{b} \right] dx \, dy$$

$$- \frac{D}{2} \int_0^b \int_0^a \left[\sum_{m=1}^{\infty} \sum_{n=1}^{\infty} A_{mn} \left(\frac{n \pi}{b} \right)^2 \sin \frac{m \pi x}{a} \sin \frac{n \pi y}{b} \right]^2 dx \, dy.$$

Because of the properties of orthogonality

$$\int_0^a \sin \frac{m \pi x}{a} \sin \frac{m' \pi x}{a} \, dx = 0 \qquad \text{for } m \neq m'$$

$$= \frac{a}{2} \qquad \text{for } m = m'$$

and

$$\int_{0}^{b} \sin \frac{m\pi y}{b} \sin \frac{n'\pi y}{b} \, dy = 0 \qquad \text{for } n \neq n'$$

$$= \frac{b}{2} \qquad \text{for } n = n'$$

the work function becomes, after integration

$$W = p_0 \sum_{m=1}^{\infty} \sum_{n=1}^{\infty} A_{mn} \frac{a}{m\pi} \frac{b}{n\pi} (\cos m\pi - 1)(\cos n\pi - 1)$$

$$- \frac{D}{2} \sum_{m=1}^{\infty} \sum_{n=1}^{\infty} A_{mn}^2 \left(\frac{m\pi}{a}\right)^4 \frac{a}{2} \frac{b}{2} - D \sum_{m=1}^{\infty} \sum_{n=1}^{\infty} A_{mn}^2 \left(\frac{m\pi}{a}\right)^2 \left(\frac{n\pi}{a}\right)^2 \frac{a}{2} \frac{b}{2}$$

$$- \frac{D}{2} \sum_{m=1}^{\infty} \sum_{n=1}^{\infty} A_{mn}^2 \left(\frac{n\pi}{b}\right)^4 \frac{a}{2} \frac{b}{2} \, .$$

To satisfy the condition of a stationary value of W

$$\frac{\partial W}{\partial A_{mn}} = 0$$

or

$$p_0 \frac{a}{m\pi} \frac{b}{n\pi} (\cos m\pi - 1)(\cos n\pi - 1) - DA_{mn} \left(\frac{m\pi}{a}\right)^4 \frac{ab}{4}$$

$$- 2DA_{mn} \left(\frac{m\pi}{a}\right)^2 \left(\frac{n\pi}{b}\right)^2 \frac{ab}{4} - DA_{mn} \left(\frac{n\pi}{b}\right)^4 \frac{ab}{4} = 0.$$

The coefficients A_{mn} can now be expressed as follows.

$$A_{mn} = \frac{p_0 \dfrac{a}{m\pi} \dfrac{b}{n\pi} (\cos m\pi - 1)(\cos n\pi - 1)}{D \dfrac{ab}{4} \left[\left(\dfrac{m\pi}{a}\right)^4 + 2\left(\dfrac{m\pi}{a}\right)^2 \left(\dfrac{n\pi}{b}\right)^2 + \left(\dfrac{n\pi}{b}\right)^4\right]}$$

The deflection becomes

$$w = \sum_{m=1}^{\infty} \sum_{n=1}^{\infty} \frac{p_0 \dfrac{a}{m\pi} \dfrac{b}{n\pi} (\cos m\pi - 1)(\cos n\pi - 1)}{D \dfrac{ab}{4} \left[\left(\dfrac{m\pi}{a}\right)^4 + 2\left(\dfrac{m\pi}{a}\right)^2 \left(\dfrac{n\pi}{b}\right)^2 + \left(\dfrac{n\pi}{b}\right)^2\right]}$$

$$\times \sin \frac{m\pi x}{a} \sin \frac{n\pi y}{b}$$

or

$$w = \frac{16p_0}{D\pi^6} \sum_{\substack{m=1 \\ odd}}^{\infty} \sum_{\substack{n=1 \\ odd}}^{\infty} \frac{1}{mn\left[\left(\frac{m}{a}\right)^2 + \left(\frac{n}{b}\right)^2\right]^2} \sin\frac{m\pi x}{a} \sin\frac{n\pi y}{b}.$$

We may refer to Chapter 1 for the expressions that give displacements, strains, stresses, and stress resultants in terms of the deflection of the middle surface.

This series for the deflection converges rapidly. For example, in the case of the deflection at the center of a square plate, the first term of the series gives a value which deviates approximately 2.5 percent from the value to which the series converges.

Example

Determine the deflection of a rectangular plate simply supported on two opposite edges, clamped on one edge and free on one edge, and subjected to a uniform lateral load, $p(x,y) = p_0$, as shown in Fig. 7.4. The work function, obtained from Eq. (7.16), is

$$W = \int_{A_1} p_0 w \, dx \, dy - \frac{D}{2} \int_{A} \left[\left(\frac{\partial^2 w}{\partial x^2}\right)^2 + \left(\frac{\partial^2 w}{\partial y^2}\right)^2 + 2(1 - \nu)\left(\frac{\partial^2 w}{\partial x \, \partial y}\right)^2 \right.$$
$$\left. + 2\nu \frac{\partial^2 w}{\partial x^2} \frac{\partial^2 w}{\partial y^2}\right] dx \, dy.$$

If the deflection is taken in the form of an infinite series or a series of even a few terms, the integration process is unwieldy unless these deflection functions and their derivatives possess the properties of

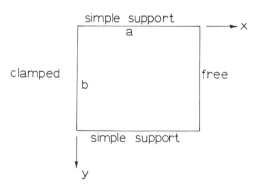

Fig. 7.4. Plate with two opposite edges simply supported, one edge clamped, one edge free, with uniform lateral load p_0.

orthogonality. For the plate considered in this example, it is difficult to find such a set of orthogonal functions that also satisfy the forced boundary conditions. Thus, an approximate solution is considered by taking the deflection in the form

$$w = A \left(\frac{x}{a} \right)^2 \sin \frac{\pi y}{b}$$

where A is a coefficient to be determined. This form satisfies the following required forced boundary conditions for this problem.

$$w = 0 \qquad 0 \leqq x \leqq a, \quad y = 0$$

$$w = 0 \qquad 0 \leqq x \leqq a, \quad y = b$$

$$w = 0 \qquad x = 0, \quad 0 \leqq y \leqq b$$

$$\frac{\partial w}{\partial x} = 0 \qquad x = 0, \quad 0 \leqq y \leqq b$$

Notice that the assumed form of the deflection does not violate the conditions of $w \neq 0$ and $\partial w / \partial x \neq 0$ along the free edge, and $\partial w / \partial y \neq 0$ along the simply supported edges.

Before we proceed to determine this coefficient, we shall simplify the expressions for the work function and the deflection function with the following transformations.

$$x = a \zeta$$

$$y = b \eta$$

Thus, we have

$$W = ab \int_{A_1} p_0 w \, d\zeta \, d\eta - \frac{Db}{2a^3} \int_A \left[\left(\frac{\partial^2 w}{\partial \zeta^2} \right)^2 + \left(\frac{a}{b} \right)^4 \left(\frac{\partial^2 w}{\partial \eta^2} \right)^2 \right.$$

$$\left. + 2(1 - \nu) \left(\frac{a}{b} \right)^2 \left(\frac{\partial^2 w}{\partial \zeta \partial \eta} \right)^2 + 2\nu \left(\frac{a}{b} \right)^2 \frac{\partial^2 w}{\partial \zeta^2} \frac{\partial^2 w}{\partial \eta^2} \right] d\zeta \, d\eta$$

and

$$w = A \zeta^2 \sin \pi \eta.$$

Upon substitution of the expression for the deflection into the work function, the following expression is obtained.

$$W = p_0\,ab \int_0^1 \int_0^1 A\,\zeta^2 \sin \pi\eta\,d\zeta\,d\eta \;-\; \frac{Db}{2a^3} \int_0^1 \int_0^1 \bigg[(2A \sin \pi\eta)^2$$

$$+ \left(\frac{a}{b}\right)^4 (-\pi^2 A\,\zeta^2 \sin \pi\eta)^2 + 2(1 - \nu)\left(\frac{a}{b}\right)^2 (2\pi A\,\zeta \cos \pi\eta)^2$$

$$+ 2\nu\left(\frac{a}{b}\right)^2 (2A \sin \pi\eta)(-\pi^2 A\,\zeta^2 \sin \pi\eta) \bigg] d\zeta\,d\eta$$

To satisfy the condition of a stationary value of W

$$\frac{\partial W}{\partial A} = 0$$

or

$$p_0 ab \int_0^1 \int_0^1 \zeta^2 \sin \pi\eta\,d\zeta\,d\eta \;-\; \frac{Db}{2a^3} \int_0^1 \int_0^1 \bigg[8A \sin^2 \pi\eta$$

$$+ 2\left(\frac{a}{b}\right)^4 \pi^4 A\,\zeta^4 \sin^2 \pi\eta + 16(1 - \nu)\left(\frac{a}{b}\right)^2 \pi^2 A\,\zeta^2 \cos^2 \pi\eta$$

$$- 8\nu\left(\frac{a}{b}\right)^2 A\pi^2 \zeta^2 \sin^2 \pi\eta \bigg] d\zeta\,d\eta = 0.$$

After integration, the expression for the coefficient A becomes

$$A = \frac{a^4 p_0}{D} \; \frac{2}{3\pi \left[2 + \dfrac{\pi^4}{10}\left(\dfrac{a}{b}\right)^4 + \dfrac{4}{3}\pi^2(1 - \nu)\left(\dfrac{a}{b}\right)^2 - \dfrac{2}{3}\pi^2\nu\left(\dfrac{a}{b}\right)^2 \right]}$$

and the expression for the deflection is

$$w = \frac{a^4 p_0}{D} \; \frac{2}{3\pi \left[2 + \dfrac{\pi^4}{10}\left(\dfrac{a}{b}\right)^4 + \dfrac{4}{3}\pi^2(1 - \nu)\left(\dfrac{a}{b}\right)^2 - \dfrac{2}{3}\pi^2\nu\left(\dfrac{a}{b}\right)^2 \right]}$$

$$\times \left(\frac{x}{a}\right)^2 \sin \frac{\pi y}{b}.$$

If we evaluate the deflection for

$$\frac{a}{b} = 1, \qquad \nu = .3, \qquad x = a, \qquad \text{and} \qquad y = \frac{b}{2}$$

we obtain

$$w = .01118 \, \frac{p_0\,a^4}{D}$$

which deviates from an exact solution[1] by approximately one percent. We may conclude that the method of Ritz is a powerful tool for obtaining approximate solutions to many difficult problems. The accuracy of the solution depends largely on the judgment used in choosing the form of the deflection functions.

7.6 METHOD OF RITZ WITH LAGRANGE MULTIPLIERS

In many problems involving the deflections of plates for which the method of Ritz is used, it is indeed difficult to construct a series of assumed functions for the deflection that satisfies all the boundary conditions, especially if these functions are to possess the properties of orthogonality. It is possible to avoid this difficulty by the introduction of Lagrange multipliers to enforce boundary conditions or constraints not satisfied by the assumed series of functions. Let us investigate the process of determining the stationary value of a work function, W, which does not meet all of the forced boundary conditions.

Suppose the deflection is taken in the form

$$w = C_1 \psi_1(x,y) + C_2 \psi_2(x,y) + C_3 \psi_3(x,y). \qquad (7.34)$$

The work function now becomes, after integration, a function of the three unknown coefficients, C_1, C_2, and C_3.

$$W = W(C_1, C_2, C_3) \qquad (7.35)$$

Furthermore, let us suppose that the assumed deflection, given by Eq. (7.34), satisfies all but two of the forced boundary conditions. These two boundary conditions or constraints, which may be expressed in terms of the three unknown coefficients of the deflection function, are

$$g_1(C_1, C_2, C_3) = 0 \qquad (7.36a)$$

$$g_2(C_1, C_2, C_3) = 0. \qquad (7.36b)$$

Our problem involves determining the stationary value of a function of three unknowns which is subjected to two functionally independent[19] constraints. The procedure may easily be extended to a function of an arbitrary number of unknowns subjected to an arbitrary number of functionally independent constraints (providing the number of constraints is less than the number of unknowns). However, clarity is best preserved at no expense of rigor by considering a function of three unknowns and two constraints.

According to Eq. (7.30) we may write

$$\frac{\partial W}{\partial C_1} \delta C_1 + \frac{\partial W}{\partial C_2} \delta C_2 + \frac{\partial W}{\partial C_3} \delta C_3 = 0. \tag{7.37}$$

If we take the first variation of each of the two constraints according to the rules of operation for the δ operator, we have

$$\frac{\partial g_1}{\partial C_1} \delta C_1 + \frac{\partial g_1}{\partial C_2} \delta C_2 + \frac{\partial g_1}{\partial C_3} \delta C_3 = 0 \tag{7.38a}$$

$$\frac{\partial g_2}{\partial C_1} \delta C_1 + \frac{\partial g_2}{\partial C_2} \delta C_2 + \frac{\partial g_2}{\partial C_3} \delta C_3 = 0. \tag{7.38b}$$

If we multiply Eq. (7.38a) by the constant λ_1, and Eq. (7.38b) by the constant λ_2, and add the results to Eq. (7.37), we obtain the following expression after rearranging terms.

$$\left(\frac{\partial W}{\partial C_1} + \frac{\partial g_1}{\partial C_1} \lambda_1 + \frac{\partial g_2}{\partial C_1} \lambda_2 \right) \delta C_1$$

$$+ \left(\frac{\partial W}{\partial C_2} + \frac{\partial g_1}{\partial C_2} \lambda_1 + \frac{\partial g_2}{\partial C_2} \lambda_2 \right) \delta C_2$$

$$+ \left(\frac{\partial W}{\partial C_3} + \frac{\partial g_1}{\partial C_3} \lambda_1 + \frac{\partial g_2}{\partial C_3} \lambda_2 \right) \delta C_3 = 0 \tag{7.39}$$

The constants λ_1 and λ_2 are called Lagrange multipliers. Since we have three unknown coefficients, C_1, C_2, and C_3, and two independent equations or expressions of constraint, g_1 and g_2, only one of the unknown coefficients is independent. The arbitrary Lagrange multipliers are chosen such that the coefficients of two of the virtual changes of Eq. (7.39) vanish. Let these two be δC_1 and δC_2. The remaining virtual change, δC_3, may now be considered the independent one. Since δC_3 is independent, it may be arbitrarily assigned so that its coefficient in Eq. (7.39) may also vanish. Thus

$$\frac{\partial W}{\partial C_1} + \frac{\partial g_1}{\partial C_1} \lambda_1 + \frac{\partial g_2}{\partial C_1} \lambda_2 = 0 \tag{7.40a}$$

$$\frac{\partial W}{\partial C_2} + \frac{\partial g_1}{\partial C_2} \lambda_1 + \frac{\partial g_2}{\partial C_2} \lambda_2 = 0 \tag{7.40b}$$

$$\frac{\partial W}{\partial C_3} + \frac{\partial g_1}{\partial C_3} \lambda_1 + \frac{\partial g_2}{\partial C_3} \lambda_2 = 0. \tag{7.40c}$$

If Eqs. (7.40a) and (7.40b) are not independent equations in the parameters λ_1 and λ_2, then[19]

$$\begin{vmatrix} \dfrac{\partial g_1}{\partial C_1} & \dfrac{\partial g_2}{\partial C_1} \\[2ex] \dfrac{\partial g_1}{\partial C_2} & \dfrac{\partial g_2}{\partial C_2} \end{vmatrix} = 0. \tag{7.41}$$

Since the determinant of a matrix is equal to the determinant of the transpose of the matrix, Eq. (7.41) may be written as

$$\begin{vmatrix} \dfrac{\partial g_1}{\partial C_1} & \dfrac{\partial g_1}{\partial C_2} \\[2ex] \dfrac{\partial g_2}{\partial C_1} & \dfrac{\partial g_2}{\partial C_2} \end{vmatrix} = 0 \tag{7.42}$$

or

$$\frac{\partial (g_1, g_2)}{\partial (C_1, C_2)} = 0. \tag{7.43}$$

Equation (7.43), which shows the Jacobian of the constraints g_1 and g_2 set equal to zero, is the condition for functional dependence of g_1 and g_2. Since these constraints are not functionally dependent, the Jacobian cannot vanish; therefore, Eqs. (7.40a) and (7.40b) are necessarily independent equations in λ_1 and λ_2. Equations (7.36) and (7.40) form a set of five independent equations which we must solve for the five unknowns, C_1, C_2, C_3, λ_1, and λ_2. These five equations may also be generated by defining a modified work function, W^*, as follows.

$$W^* = W(C_1, C_2, C_3) + \lambda_1 g_1(C_1, C_2, C_3) + \lambda_2 g_2(C_1, C_2, C_3) \tag{7.44a}$$

or

$$W^* = W^*(C_1, C_2, C_3, \lambda_1, \lambda_2) \tag{7.44b}$$

A stationary value of the function W^* requires that

$$\frac{\partial W^*}{\partial C_1} = 0 = \frac{\partial W}{\partial C_1} + \frac{\partial g_1}{\partial C_1} \lambda_1 + \frac{\partial g_2}{\partial C_1} \lambda_2 \tag{7.45a}$$

$$\frac{\partial W^*}{\partial C_2} = 0 = \frac{\partial W}{\partial C_2} + \frac{\partial g_1}{\partial C_2} \lambda_1 + \frac{\partial g_2}{\partial C_2} \lambda_2 \qquad (7.45b)$$

$$\frac{\partial W^*}{\partial C_3} = 0 = \frac{\partial W}{\partial C_3} + \frac{\partial g_1}{\partial C_3} \lambda_1 + \frac{\partial g_2}{\partial C_3} \lambda_2 \qquad (7.45c)$$

$$\frac{\partial W^*}{\partial \lambda_1} = 0 = g_1 \qquad (7.45d)$$

$$\frac{\partial W^*}{\partial \lambda_2} = 0 = g_2. \qquad (7.45e)$$

These equations are identical to Eqs. (7.36) and (7.40); thus, they form a set of five independent equations in the five unknowns, C_1, C_2, C_3, λ_1, and λ_2. Once these five unknowns are determined for a particular problem, the problem is solved.

A Lagrange multiplier has a specific physical significance. It represents the reaction necessary to maintain the constraint to which it is associated. The terms $\lambda_1 g_1$ and $\lambda_2 g_2$ of the modified work function of Eq. (7.44a) represent the work of these reactions.

Some of the advantages of this method over the conventional Ritz method are

1. Since the assumed deflection function is not required to satisfy all of the forced boundary conditions, orthogonal functions may be selected, which simplifies the integration procedures.

2. Problems with unusual constraints may be solved.

3. Constraint reactions may be determined as part of the solution.

Example

Determine the deflection of a rectangular plate simply supported on three edges, clamped on the fourth edge, and subjected to a uniform lateral load, $p(x,y) = p_0$, as shown in Fig. 7.5.

The work function, obtained from Eq. (7.22), is

$$W = \int_{A_1} p_0 w \, dx \, dy - \frac{D}{2} \int_A \left[\left(\frac{\partial^2 w}{\partial x^2} \right)^2 + 2 \frac{\partial^2 w}{\partial x^2} \frac{\partial^2 w}{\partial y^2} + \left(\frac{\partial^2 w}{\partial y^2} \right)^2 \right] dx \, dy$$

$$+ D(1 - \nu) \oint_C \frac{\partial w}{\partial x} \frac{\partial^2 w}{\partial x \, \partial y} \, dx + D(1 - \nu) \oint_C \frac{\partial w}{\partial x} \frac{\partial^2 w}{\partial y^2} \, dy.$$

For the plate supported as shown in Fig. 7.5, we see that the last two integrals must vanish; thus, the work function reduces to the following expression.

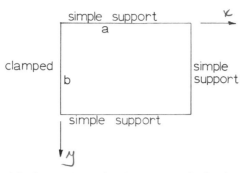

Fig. 7.5. Plate with three edges simply supported, the fourth edge clamped, and uniform lateral load p_0

$$W = \int_{A_1} p_0 w \, dx \, dy - \frac{D}{2} \int_A \left[\left(\frac{\partial^2 w}{\partial x^2} \right)^2 + 2 \frac{\partial^2 w}{\partial x^2} \frac{\partial^2 w}{\partial y^2} + \left(\frac{\partial^2 w}{\partial y^2} \right)^2 \right] dx \, dy$$

The deflection of the plate is taken in the form of a double trigonometric sine series as follows.

$$w = \sum_{m=1}^{\infty} \sum_{n=1}^{\infty} A_{mn} \sin \frac{m\pi x}{a} \sin \frac{n\pi y}{b}$$

This form for the deflection satisfies the forced boundary conditions of zero deflection on the three simply supported edges and the clamped edge. However, it does not satisfy the forced boundary condition of zero slope on the clamped edge:

$$\frac{\partial w}{\partial x} = 0 \qquad \text{at} \qquad x = 0, \quad 0 \leqq y \leqq b$$

or plugging in

$$\sum_{m=1}^{\infty} \sum_{n=1}^{\infty} m A_{mn} \sin \frac{n\pi y}{b} = 0$$

or expanding;

$$\sin \frac{\pi y}{b} \sum_{m=1}^{\infty} m A_{m1} + \sin \frac{2\pi y}{b} \sum_{m=1}^{\infty} m A_{m2} + \ldots = 0.$$

Since y may be arbitrarily assigned, this constraint equation is satisfied as follows. *(factor out of term)*

$$\sum_{m=1}^{\infty} m\, A_{mj} = 0 \qquad j = 1, 2, \ldots, \infty$$

(mathematically ??)
We see now that we actually have an infinite number of constraints rather than just one constraint, since we are satisfying a condition along an edge rather than at a point. The modified work function is

$$W^* = W + \lambda_1 \sum_{m=1}^{\infty} m\, A_{m1} + \lambda_2 \sum_{m=1}^{\infty} m\, A_{m2} + \ldots$$

or

$$W^* = W + \sum_{m=1}^{\infty} \sum_{n=1}^{\infty} \lambda_n\, m\, A_{mn}.$$

Since the constraint must be satisfied along the entire edge, we see that an infinite number of Lagrange multipliers is required.

If we substitute the expression for the deflection into the modified work function and perform the necessary integrations, taking into account the properties of orthogonality, we obtain

$$W^* = \frac{p_0\, ab}{\pi^2} \sum_{m=1}^{\infty} \sum_{n=1}^{\infty} \frac{(\cos m\pi - 1)(\cos n\pi - 1)}{mn} A_{mn}$$

$$- \frac{Dab\,\pi^4}{8} \sum_{m=1}^{\infty} \sum_{n=1}^{\infty} \left[\left(\frac{m}{a}\right)^2 + \left(\frac{n}{b}\right)^2 \right]^2 A_{mn}^2 + \sum_{m=1}^{\infty} \sum_{n=1}^{\infty} \lambda_n\, m\, A_{mn}.$$

A stationary value of the function W^* requires that

$$\frac{\partial W^*}{\partial A_{mn}} = 0 \qquad \text{for each } A_{mn}$$

and

$$\frac{\partial W^*}{\partial \lambda_n} = 0 \qquad \text{for each } \lambda_n.$$

Performing these partial derivatives, we obtain the following expressions.

$$\frac{\partial W^*}{\partial A_{mn}} = 0 = \frac{p_0 \, ab}{\pi^2} \frac{(\cos m\pi - 1)(\cos n\pi - 1)}{mn}$$
$$- \frac{Dab \, \pi^4}{4} \left[\left(\frac{m}{a}\right)^2 + \left(\frac{n}{b}\right)^2 \right]^2 A_{mn} + \lambda_n \, m$$

or

$$A_{mn} = \frac{\dfrac{4p_0 \, ab}{\pi^2} \dfrac{(\cos m\pi - 1)(\cos n\pi - 1)}{mn} + 4m \, \lambda_n}{Dab \, \pi^4 \left[\left(\dfrac{m}{a}\right)^2 + \left(\dfrac{n}{b}\right)^2 \right]^2}$$

and

$$\frac{\partial W^*}{\partial \lambda_n} = 0 = \sum_{m=1}^{\infty} m \, A_{mn} \qquad n = 1, 2, \ldots$$

If we substitute the expression for A_{mn} into the equation $0 = \sum_{m=1}^{\infty} m A_{mn}$, then solve for λ_n, we have

$$0 = \frac{4p_0 \, a^4}{D\pi^6} \sum_{m=1}^{\infty} \frac{(\cos m\pi - 1)(\cos n\pi - 1)}{n \left[m^2 + n^2 \left(\dfrac{a}{b}\right)^2 \right]^2} + \frac{4a^3 \lambda_n}{Db \, \pi^4} \sum_{m=1}^{\infty} \frac{m^2}{\left[m^2 + n^2 \left(\dfrac{a}{b}\right)^2 \right]^2}$$

or

$$\lambda_n = -\frac{p_0 \, ab}{\pi^2} \frac{\displaystyle\sum_{m=1}^{\infty} \frac{(\cos m\pi - 1)(\cos n\pi - 1)}{n \left[m^2 + n^2 \left(\dfrac{a}{b}\right)^2 \right]^2}}{\displaystyle\sum_{m=1}^{\infty} \frac{m^2}{\left[m^2 + n^2 \left(\dfrac{a}{b}\right)^2 \right]^2}}$$

or

$$\lambda_n = -\frac{p_0 \, ab}{\pi^2} B_n.$$

The Lagrange multiplier represents the moment per unit length necessary to force zero slope along the clamped edge. The negative sign indicates that the moment is in a direction opposite to that which is considered a positive direction of rotation or slope.

For a particular value of n, B_n and thus λ_n can be obtained by summing over m in the series in the numerator and the series in the denominator independently and dividing. When λ_n is known, A_{mn} can be determined by direct substitution.

Since A_{mn} is now completely defined, the deflection function may now be expressed as follows.

$$w = \frac{4p_0 a^4}{D\pi^6} \left\{ \sum_{m=1}^{\infty} \sum_{n=1}^{\infty} \frac{(\cos m\pi - 1)(\cos n\pi - 1)}{mn\left[m^2 + n^2\left(\frac{a}{b}\right)^2\right]^2} \sin\frac{m\pi x}{a} \sin\frac{n\pi y}{b} \right.$$

$$\left. - \sum_{m=1}^{\infty} \sum_{n=1}^{\infty} \frac{mB_m}{\left[m^2 + n^2\left(\frac{a}{b}\right)^2\right]^2} \sin\frac{m\pi x}{a} \sin\frac{n\pi y}{b} \right\}$$

If we evaluate this expression for $x = a/2$, $y = b/2$, and $a/b = 1$, we obtain the deflection at the midpoint of a square plate as

$$w = .002785 \, \frac{p_0 a^4}{D}.$$

Example

Determine the deflection of a rectangular plate simply supported on the four outer edges with a point support at the center and subjected to a uniform lateral load, $p(x,y) = p_0$, as shown in Fig. 7.6.

The work function, obtained from Eq. (7.22) is

$$W = \int_{A_1} p_0 w \, dx \, dy - \frac{D}{2} \int_A \left[\left(\frac{\partial^2 w}{\partial x^2}\right)^2 + 2\frac{\partial^2 w}{\partial x^2}\frac{\partial^2 w}{\partial y^2} + \left(\frac{\partial^2 w}{\partial y^2}\right)^2\right] dx \, dy$$

$$+ D(1 - \nu) \oint_C \frac{\partial w}{\partial x}\frac{\partial^2 w}{\partial x \, \partial y} \, dx + D(1 - \nu) \oint_C \frac{\partial w}{\partial x}\frac{\partial^2 w}{\partial y^2} \, dy.$$

For a plate simply supported along its edges, we see that the last two integrals must vanish; thus, the work function reduces to the following expression.

$$W = \int_{A_1} p_0 w \, dx \, dy - \frac{D}{2} \int_A \left[\left(\frac{\partial^2 w}{\partial x^2}\right)^2 + 2\frac{\partial^2 w}{\partial x^2}\frac{\partial^2 w}{\partial y^2} + \left(\frac{\partial^2 w}{\partial y^2}\right)^2\right] dx \, dy$$

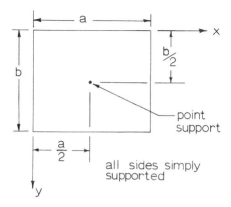

Fig. 7.6. Simply supported plate with point support at the center and uniform lateral load p_0

The deflection of the plate is taken in the form of a double trigonometric sine series as follows.

$$w = \sum_{m=1}^{\infty} \sum_{n=1}^{\infty} A_{mn} \sin \frac{m\pi x}{a} \sin \frac{n\pi y}{b}$$

This form for the deflection satisfies the forced boundary conditions along all the edges of the plate; however, the constraint of zero deflection at the center is not satisfied. This constraint is

$$w = 0 \qquad \text{at} \qquad x = \frac{a}{2}, \quad y = \frac{b}{2}$$

or

$$\sum_{m=1}^{\infty} \sum_{n=1}^{\infty} A_{mn} \sin \frac{m\pi}{2} \sin \frac{n\pi}{2} = 0.$$

The modified work function is

$$W^* = W + \lambda \sum_{m=1}^{\infty} \sum_{n=1}^{\infty} A_{mn} \sin \frac{m\pi}{2} \sin \frac{n\pi}{2}.$$

If we substitute the expression for the deflection into the modified work function and perform the necessary integrations, taking into account the properties of orthogonality, we obtain

$$W^* = \frac{p_0\,ab}{\pi^2} \sum_{m=1}^{\infty} \sum_{n=1}^{\infty} \frac{(\cos m\pi - 1)(\cos n\pi - 1)}{mn}\, A_{mn}$$

$$-\frac{Dab\,\pi^4}{8} \sum_{m=1}^{\infty} \sum_{n=1}^{\infty} \left[\left(\frac{m}{a}\right)^2 + \left(\frac{n}{b}\right)^2\right]^2 A_{mn}^2$$

$$+ \lambda \sum_{m=1}^{\infty} \sum_{n=1}^{\infty} A_{mn} \sin\frac{m\pi}{2} \sin\frac{n\pi}{2}.$$

A stationary value of the work function W^* requires that

$$\frac{\partial W^*}{\partial A_{mn}} = 0 \qquad \text{for each } A_{mn}$$

and

$$\frac{\partial W^*}{\partial \lambda} = 0.$$

If we perform these partial derivatives, we obtain

$$\frac{\partial W^*}{\partial A_{mn}} = 0 = \frac{p_0\,ab}{\pi^2}\,\frac{(\cos m\pi - 1)(\cos n\pi - 1)}{mn}$$

$$-\frac{Dab\,\pi^4}{4}\left[\left(\frac{m}{a}\right)^2 + \left(\frac{n}{b}\right)^2\right]^2 A_{mn} + \lambda \sin\frac{m\pi}{2}\sin\frac{n\pi}{2}$$

or

$$A_{mn} = \frac{4p_0\,a^4}{D\pi^6}\,\frac{(\cos m\pi - 1)(\cos n\pi - 1)}{mn\left[m^2 + n^2\left(\frac{a}{b}\right)^2\right]^2} + \frac{4a^3\,\lambda \sin\dfrac{m\pi}{2}\sin\dfrac{n\pi}{2}}{D\pi^4 b\left[m^2 + n^2\left(\frac{a}{b}\right)^2\right]^2}$$

and

$$\frac{\partial W^*}{\partial \lambda} = 0 = \sum_{m=1}^{\infty} \sum_{n=1}^{\infty} A_{mn}\sin\frac{m\pi}{2}\sin\frac{n\pi}{2}.$$

If we substitute the expression for A_{mn} into the equation

$$0 = \sum_{m=1}^{\infty} \sum_{n=1}^{\infty} A_{mn}\sin\frac{m\pi}{2}\sin\frac{n\pi}{2}, \text{ then solve for } \lambda, \text{ we have}$$

$$0 = \frac{4p_0 a^4}{D\pi^6} \sum_{m=1}^{\infty} \sum_{n=1}^{\infty} \frac{(\cos m\pi - 1)(\cos n\pi - 1)}{mn\left[m^2 + n^2\left(\dfrac{a}{b}\right)^2\right]^2} \sin \frac{m\pi}{2} \sin \frac{n\pi}{2}$$

$$+ \frac{4a^3 \lambda}{D\pi^4 b} \sum_{m=1}^{\infty} \sum_{n=1}^{\infty} \frac{\left(\sin \dfrac{m\pi}{2} \sin \dfrac{n\pi}{2}\right)^2}{\left[m^2 + n^2\left(\dfrac{a}{b}\right)^2\right]^2} \; \sin \frac{m\pi}{2} \sin \frac{n\pi}{2}$$

or

$$\lambda = -\frac{p_0 ab}{\pi^2} \; \frac{\displaystyle\sum_{m=1}^{\infty} \sum_{n=1}^{\infty} \frac{(\cos m\pi - 1)(\cos n\pi - 1)}{mn\left[m^2 + n^2\left(\dfrac{a}{b}\right)^2\right]^2} \sin \frac{m\pi}{2} \sin \frac{n\pi}{2}}{\displaystyle\sum_{m=1}^{\infty} \sum_{n=1}^{\infty} \frac{\left(\sin \dfrac{m\pi}{2} \sin \dfrac{n\pi}{2}\right)^2}{\left[m^2 + n^2\left(\dfrac{a}{b}\right)^2\right]^2} \; \sin \frac{m\pi}{2} \sin \frac{n\pi}{2}}$$

or

$$\lambda = -\frac{p_0 ab}{\pi^2} B.$$

For this problem, the Lagrange multiplier represents the force necessary to maintain zero deflection at the point support. The negative sign
indicates that the force is in a direction opposite to that which is considered a positive direction of deflection.

The constant B is determined by summing the numerator and denominator separately and dividing.

Since A_{mn} is now completely defined, the deflection function may now
be expressed as follows.

$$w = \frac{4p_0 a^4}{D\pi^6} \left\{ \sum_{m=1}^{\infty} \sum_{n=1}^{\infty} \frac{(\cos m\pi - 1)(\cos n\pi - 1)}{mn\left[m^2 + n^2\left(\dfrac{a}{b}\right)^2\right]^2} \sin \frac{m\pi x}{a} \sin \frac{n\pi y}{b} \right.$$

$$\left. - B \sum_{m=1}^{\infty} \sum_{n=1}^{\infty} \frac{\sin \dfrac{m\pi}{2} \sin \dfrac{n\pi}{2}}{\left[m^2 + n^2\left(\dfrac{a}{b}\right)^2\right]^2} \sin \frac{m\pi x}{a} \sin \frac{n\pi y}{b} \right\}$$

Example[31]

In the solution of the problem in Fig. 7.7, the forcing function is assumed to be a uniformly distributed load. The deflected surface is

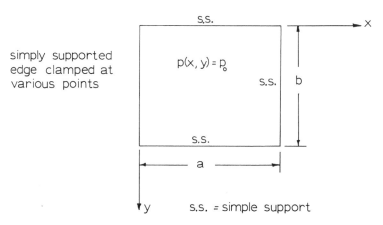

Fig. 7.7

approximated by a function consisting of an infinite sine series on y and a fourth-order polynomial on x.

$$w = \sum_{m=1}^{4} r^m \sum_{n=1}^{\infty} A_{mn} \sin \frac{n\pi y}{b}, \quad \text{where } r = \frac{x}{a}$$

A double sine series is not used, since the constraint of zero slope at a point due to a point clamped along the edge results in a divergent series. The polynomial on x is of the fourth order to insure that there are a greater number of unknown constants in the assumed deflection series than the number of constraints.

As in the previous examples, the work function is

$$W = \int_{A_1} p_0 w \, dx \, dy - \frac{D}{2} \int_A \left[\left(\frac{\partial^2 w}{\partial x^2} \right)^2 + 2 \frac{\partial^2 w}{\partial x^2} \frac{\partial^2 w}{\partial y^2} + \left(\frac{\partial^2 w}{\partial y^2} \right)^2 \right] dx \, dy.$$

If we substitute the expression for the deflection into the work function and perform the necessary integrations, we obtain the following.

$$W = \frac{p_0 ab}{\pi} \left[\sum_{m=1}^{4} \sum_{n=1}^{\infty} \frac{1}{n(m+1)} A_{mn} (\cos n\pi - 1) \right]$$

$$-\frac{Db}{4a^3}\left[\sum_{n=1}^{\infty}\frac{1}{3}B_n^2 A_{1n}^2 - \sum_{n=1}^{\infty}2B_n\left(1-\frac{1}{4}B_n\right)A_{1n}A_{2n}\right.$$

$$-\sum_{n=1}^{\infty}2B_n\left(2-\frac{1}{5}B_n\right)A_{1n}A_{3n} - \sum_{n=1}^{\infty}2B_n\left(3-\frac{1}{6}B_n\right)A_{1n}A_{4n}$$

$$+\sum_{n=1}^{\infty}\left(4-\frac{4}{3}B_n+\frac{1}{5}B_n^2\right)A_{2n}^2 + \sum_{n=1}^{\infty}2\left(6-2B_n+\frac{1}{6}B_n^2\right)A_{2n}A_{3n}$$

$$+\sum_{n=1}^{\infty}2\left(8-\frac{14}{5}B_n+\frac{1}{7}B_n^2\right)A_{2n}A_{4n} + \sum_{n=1}^{\infty}\left(12-\frac{12}{5}B_n+\frac{1}{7}B_n^2\right)A_{3n}^2$$

$$+\sum_{n=1}^{\infty}2\left(18-3B_n+\frac{1}{8}B_n^2\right)A_{3n}A_{4n} + \sum_{n=1}^{\infty}\left(\frac{144}{5}-\frac{24}{7}B_n+\frac{1}{9}B_n^2\right)A_{4n}^2\right]$$

$$\text{where } B_n = \frac{n^2\pi^2 a^2}{b^2}$$

For a fully simply supported plate, the forced boundary conditions are

$$w = 0 \quad \text{at} \quad r = 0, \quad 0 \le y \le b$$

$$w = 0 \quad \text{at} \quad r = 1, \quad 0 \le y \le b$$

$$w = 0 \quad \text{at} \quad y = 0, \quad 0 \le r \le 1$$

$$w = 0 \quad \text{at} \quad y = b, \quad 0 \le r \le 1.$$

All of the forced boundary conditions are satisfied by the assumed deflection function except

$$w = 0 \quad \text{at} \quad r = 1, \quad 0 \le y \le b.$$

Therefore, the constraint equation is

$$w\bigg|_{r=1} = \sum_{m=1}^{4}\sum_{n=1}^{\infty}A_{mn}\sin\frac{n\pi y}{b} = 0.$$

A sufficient constraint is that

$$\sum_{m=1}^{4}A_{mn} = 0, \quad n = 1, 2, \ldots, \infty.$$

Since there are an infinite number of constraint equations, an infinite number of Lagrange multipliers must be used. Thus, the modified work function is

$$W^* = W + \sum_{n=1}^{\infty} \lambda_n \sum_{m=1}^{4} A_{mn}$$

where λ is a Lagrange multiplier.

For a stationary value of W^*, the following equations must be satisfied for each value of n.

$$\frac{\partial W^*}{\partial A_{1n}} = C_{1n}^{I} A_{1n} + C_{1n}^{II} A_{2n} + C_{1n}^{III} A_{3n} + C_{1n}^{IV} A_{4n} - \frac{1}{2} D_{1n} + t \lambda_n = 0$$

$$\frac{\partial W^*}{\partial A_{2n}} = C_{2n}^{I} A_{1n} + C_{2n}^{II} A_{2n} + C_{2n}^{III} A_{3n} + C_{2n}^{IV} A_{4n} - \frac{1}{3} D_{2n} + t \lambda_n = 0$$

$$\frac{\partial W^*}{\partial A_{3n}} = C_{3n}^{I} A_{1n} + C_{3n}^{II} A_{2n} + C_{3n}^{III} A_{3n} + C_{3n}^{IV} A_{4n} - \frac{1}{4} D_{3n} + t \lambda_n = 0$$

$$\frac{\partial W^*}{\partial A_{4n}} = C_{4n}^{I} A_{1n} + C_{4n}^{II} A_{2n} + C_{4n}^{III} A_{3n} + C_{4n}^{IV} A_{4n} - \frac{1}{5} D_{4n} + t \lambda_n = 0$$

$$\frac{\partial W^*}{\partial \lambda_n} = \sum_{m=1}^{4} A_{mn} = 0, \qquad n = 1, 2, 3, \ldots, \infty$$

where

$$C_{1n}^{I} = \frac{2}{3} B_n^2 \qquad C_{2n}^{II} = 2 \left(4 - \frac{4}{3} B_n + \frac{1}{5} B_n^2 \right)$$

$$C_{3n}^{III} = 2 \left(12 - \frac{12}{5} B_n + \frac{1}{7} B_n^2 \right) \qquad C_{4n}^{IV} = 2 \left(\frac{144}{5} - \frac{24}{7} B_n + \frac{1}{9} B_n^2 \right)$$

$$C_{1n}^{II} = C_{2n}^{I} = -2 B_n \left(1 - \frac{1}{4} B_n \right) \qquad C_{1n}^{III} = C_{3n}^{I} = -2 B_n \left(2 - \frac{1}{5} B_n \right)$$

$$C_{1n}^{IV} = C_{4n}^{I} = -2 B_n \left(3 - \frac{1}{6} B_n \right) \qquad C_{2n}^{III} = C_{3n}^{II} = 2 \left(6 - 2 B_n + \frac{1}{16} B_n^2 \right)$$

$$C_{2n}^{IV} = C_{4n}^{II} = 2 \left(8 - \frac{14}{5} B_n + \frac{1}{7} B_n^2 \right)$$

$$C_{4n}^{IV} = C_{4n}^{III} = 2 \left(18 - 3B_n + \frac{1}{8} B_n^2 \right).$$

$$D_{mn} = \frac{p_0 a^4}{D\pi} \frac{(\cos n\pi - 1)}{n}, \qquad m = 1, 2, 3, \text{ and } 4$$

$$t = \frac{4a^3}{Db}$$

Since the set of equations must hold for each value of n (specifically the constraint equation), the values of A_{1n}, A_{2n}, A_{3n}, A_{4n}, and $t \lambda_n$ can be obtained by the solution of the simultaneous set of equations for each value of n, thus enabling the deflections to be computed. However, one giant set of simultaneous equations can be written for all values of n (where n is limited to some finite number), thus enabling all the deflection series coefficients and Lagrange multipliers to be obtained by the solution of this one set of equations. For points clamped along an edge, constraint equations will exist that relate the deflection series coefficients for all values of n, and do not hold for each value of n, thus requiring that all of the equations be solved simultaneously.

For the solution of the problem with any or all of the edges fully clamped, or with points clamped along an edge, or any combination of these conditions, it is only necessary that the appropriate constraint equations multiplied by Lagrange multipliers be added to the modified work expression for the fully simply supported plate. The net result is that the additional constraint equations are added to the set of simultaneous equations for the fully simply supported plate, where it is observed that the matrix of coefficients of the set of simultaneous equations is symmetrical.

The following is an example derivation of the additional constraint equations and the resulting modified work expression for a finite number of points clamped along the edge at x = 0.

For one point, the additional constraint equation is

$$\left. \frac{\partial w}{\partial x} \right|_{\substack{x = 0 \\ y = b/2}} = \sum_{n=1}^{\infty} A_{1n} \sin \frac{n\pi}{2} = 0.$$

For four points, the additional constraint equations are

$$\left. \frac{\partial w}{\partial x} \right|_{\substack{x = 0 \\ y = b/5}} = \sum_{n=1}^{\infty} A_{1n} \sin \frac{n\pi}{5} = 0$$

$$\left. \frac{\partial w}{\partial x} \right|_{\substack{x\ =\ 0 \\ y\ =\ 2b/5}} = \sum_{n=1}^{\infty} A_{1n} \sin \frac{2n\pi}{5} = 0$$

$$\left. \frac{\partial w}{\partial x} \right|_{\substack{x\ =\ 0 \\ y\ =\ 3b/5}} = \sum_{n=1}^{\infty} A_{1n} \sin \frac{3n\pi}{5} = 0$$

and

$$\left. \frac{\partial w}{\partial x} \right|_{\substack{x\ =\ 0 \\ y\ =\ 4b/5}} = \sum_{n=1}^{\infty} A_{1n} \sin \frac{4n\pi}{5} = 0.$$

As an example, the modified work expression for the case of four points clamped is

$$W^* = W + \sum_{n=1}^{\infty} \lambda_n \sum_{m=1}^{4} A_{mn} + \mu_1 \sum_{n=1}^{\infty} A_{1n} \sin \frac{n\pi}{5}$$

$$+ \mu_2 \sum_{n=1}^{\infty} A_{1n} \sin \frac{2n\pi}{5} + \mu_3 \sum_{n=1}^{\infty} A_{1n} \sin \frac{3n\pi}{5} + \mu_4 \sum_{n=1}^{\infty} A_{1n} \sin \frac{4n\pi}{5}$$

where λ_n, μ_1, μ_2, μ_3, and μ_4 are Lagrange multipliers.

Solutions for the following six problems were obtained. These problems consist of various combinations of clamped and simply supported edges, at $y = 0$, $y = b$, and $x = a$. In addition, solutions for each of the six problems were obtained for the edge at $x = 0$ simply supported, simply supported with one, two, three, four, five, and six points clamped, and fully clamped. Solutions to all of these problems were obtained for b/a values of 0.5, 1.0, 1.5, 2.0, and 5.0.

	Edge at $y = 0$	Edge at $x = a$	Edge at $y = b$
Problem 1	Simply supported	Simply supported	Simply supported
Problem 2	Simply supported	Clamped	Simply supported
Problem 3	Clamped	Simply supported	Clamped
Problem 4	Clamped	Clamped	Clamped
Problem 5	Clamped	Simply supported	Simply supported
Problem 6	Clamped	Clamped	Simply supported

For each problem, deflections were calculated at the following points on the plate.

x-Coordinate	y-Coordinate
0.25a	0.25b
.50a	.25b
.75a	.25b
.25a	.50b
.50a	.50b
.75a	.50b
.25a	.75b
.50a	.75b
0.75a	0.75b

The limit of the summation on n was limited to a maximum of fifteen terms, with the deflections also calculated for fourteen terms, so that a convergence check could be made. Convergence to three significant figures was obtained for Problems 1, 2, 3, and 4, and convergence to two significant figures was obtained for Problems 5 and 6. It should be noted that for the fully clamped plate, with fifteen terms on n, a set of simultaneous equations of the order of 109 unknowns is obtained.

Tables 5–10 corresponding to problems 1–6, present the deflections at the center of the plate as a function of the number of points clamped on the edge at $x = 0$. This data is presented for the five b/a values as defined previously. All values of the deflection are in terms of $p_0 a^4 / D$.

TABLE 5

No. of Points Clamped on Edge x = 0	Deflection at Center of Plate for Problem 1				
	$b/a = 0.5$	$b/a = 1.0$	$b/a = 1.5$	$b/a = 2.0$	$b/a = 5.0$
0	0.000643	0.00408	0.00774	0.01014	0.01297
1	.000601	.00314	.00548	.00700	.00876
2	.000591	.00288	.00480	.00619	.01055
3	.000591	.00284	.00456	.00516	.00816
4	.000590	.00281	.00438	.00525	.00813
5	.000590	.00280	.00434	.00514	.00724
6	.000590	.00280	.00429	.00501	.00688
∞	0.000590	0.00279	0.00426	0.00489	0.00521

TABLE 6

No. of Points Clamped on Edge x = 0	Deflection at Center of Plate for Problem 2				
	$b/a = 0.5$	$b/a = 1.0$	$b/a = 1.5$	$b/a = 2.0$	$b/a = 5.0$
0	0.000590	0.00279	0.00426	0.00489	0.00521
1	.000557	.00216	.00311	.00354	.00372
2	.000552	.00200	.00275	.00319	.00468
3	.000552	.00197	.00262	.00294	.00365
4	.000551	.00196	.00254	.00276	.00376
5	.000551	.00195	.00252	.00271	.00339
6	.000551	.00195	.00251	.00267	.00325
∞	0.000551	0.00195	0.00249	0.00262	0.00260

TABLE 7

No. of Points Clamped on Edge x = 0	b/a = 0.5	b/a = 1.0	b/a = 1.5	b/a = 2.0	b/a = 5.0
		Deflection at Center of Plate for Problem 3			
0	0.000165	0.00186	0.00519	0.00830	0.01291
1	.000165	.00161	.00394	.00589	.00872
2	.000164	.00155	.00368	.00549	.01053
3	.000164	.00155	.00354	.00509	.00815
4	.000164	.00153	.00343	.00476	.00812
5	.000164	.00153	.00337	.00467	.00724
6	.000164	.00153	.00338	.00456	.00688
∞	0.000164	0.00153	0.00336	0.00446	0.00521

TABLE 8

No. of Points Clamped on Edge x = 0	b/a = 0.5	b/a = 1.0	b/a = 1.5	b/a = 2.0	b/a = 5.0
		Deflection at Center of Plate for Problem 4			
0	0.000168	0.00153	0.00336	0.00446	0.00521
1	.000168	.00132	.00255	.00326	.00372
2	.000167	.00128	.00238	.00305	.00468
3	.000167	.00127	.00230	.00283	.00365
4	.000167	.00126	.00223	.00267	.00375
5	.000167	.00126	.00222	.00263	.00338
6	.000167	.00126	.00221	.00258	.00324
∞	0.000167	0.00126	0.00220	0.00254	0.00260

TABLE 9

No. of Points Clamped on Edge x = 0	b/a = 0.5	b/a = 1.0	b/a = 1.5	b/a = 2.0	b/a = 5.0
		Deflection at Center of Plate for Problem 5			
0	0.00029	0.0027	0.0063	0.0091	0.0130
1	.00029	.0022	.0046	.0064	.0083
2	.00029	.0021	.0042	.0058	.0103
3	.00029	.0021	.0040	.0053	.0077
4	.00029	.0020	.0038	.0049	.0077
5	.00029	.0020	.0038	.0049	.0069
6	.00029	.0020	.0038	.0048	.0067
∞	0.00029	0.0020	0.0038	0.0047	0.0052

TABLE 10

No. of Points Clamped on Edge x = 0	b/a = 0.5	b/a = 1.0	b/a = 1.5	b/a = 2.0	b/a = 5.0
		Deflection at Center of Plate for Problem 6			
0	0.00029	0.0020	0.0038	0.0047	0.0052
1	.00029	.0017	.0028	.0034	.0036
2	.00029	.0016	.0025	.0031	.0047
3	.00029	.0016	.0024	.0029	.0035
4	.00029	.0016	.0024	.0027	.0036
5	.00029	.0016	.0024	.0027	.0033
6	.00029	.0016	.0023	.0026	.0032
∞	0.00029	0.0016	0.0023	0.0026	0.0026

Figures 7.8–7.13, corresponding to problems 1–6, consist of plots of the average percentage deviation of the plate deflections from a fully clamped edge as compared to a fully simply supported edge. These values are plotted as a function of the number of points clamped on the edge at x = 0. The following equation was used to calculate the percent deviation of the plate deflections from a fully clamped edge as compared to a fully simply supported edge, for each of the nine points defined previously at which the plate deflections were calculated.

$$\text{Percent deviation} = 100 \times \frac{w - w_{cl}}{w_{ss} - w_{cl}}$$

where w is the plate deflection for the specific problem, w_{cl} is the deflection for the edge at x = 0 fully clamped, and w_{ss} is the deflection for the edge at x = 0 fully simply supported. The resulting values for the nine points on the plate were then averaged to obtain the average percentage deviation.

The results of this sample problem are of considerable importance to the practicing engineer. Through the use of Figs. 7.8–7.13, the designer can determine the required number of fasteners to use on a plate to simulate a fully clamped edge.

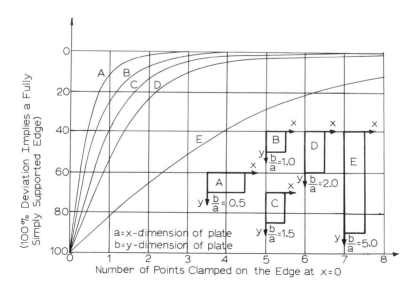

Fig. 7.8. **Average percent deviation of deflections from a fully clamped edge**

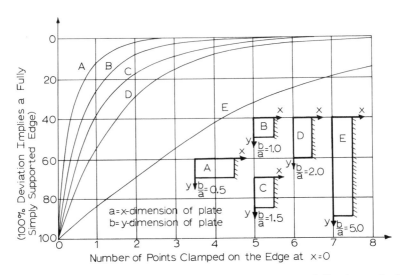

Fig. 7.9. Average percent deviation of deflections from a fully clamped edge

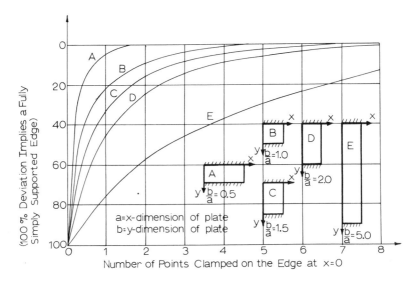

Fig. 7.10. Average percent deviation of deflections from a fully clamped edge

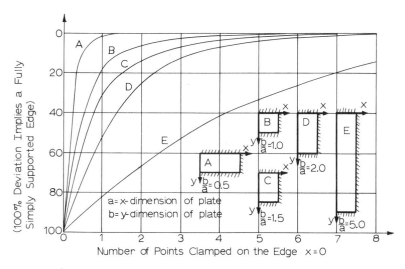

Fig. 7.11. Average percent deviation of deflections from a fully clamped edge

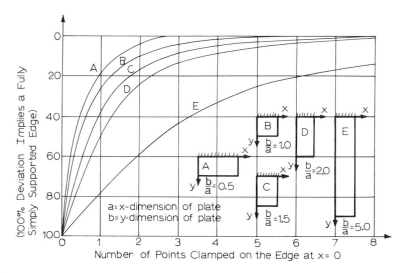

Fig. 7.12. Average percent deviation of deflections from a fully clamped edge

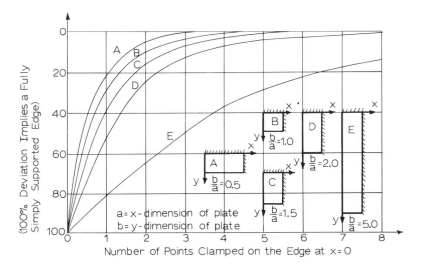

Fig. 7.13. Average percent deviation of deflections from a fully clamped edge

Problems

(29.) A simply supported rectangular plate has a concentrated load P at x = .75a and y = .5b as shown in Fig. 7.14. Use the method of Ritz to determine an expression for the deflection of the plate.

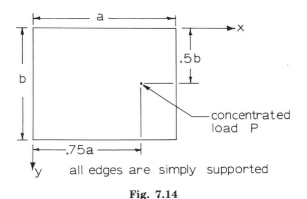

Fig. 7.14

30. A rectangular plate with the lateral load p = p_0xy is constrained as shown in Fig. 7.15. Use the method of Ritz to determine an approximate expression for the deflection of the plate.

what would assu form of assumed solution

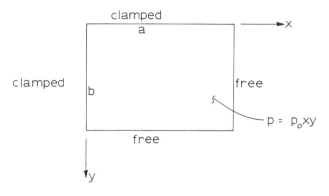

Fig. 7.15

31. A rectangular plate with a uniform load is constrained as shown in Fig. 7.16. Use the method of Ritz to determine an approximate expression for the deflection of the plate.

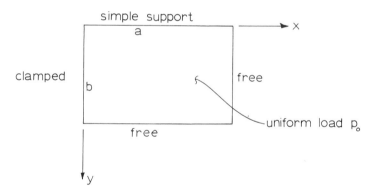

Fig. 7.16

32. Consider the plate in the second example in Section 7.5 and shown in Fig. 7.4. Use each of the following expressions to determine the lateral deflection by the method of Ritz.

a. $$w = \zeta^2 \sum_{n=1}^{\infty} A_n \sin n\pi\eta$$

b. $$w = (A_1 \zeta^2 + A_2 \zeta^3) \sin \pi\eta$$

c. $$w = (A_1 \zeta^2 + A_2 \zeta^3) \sum_{n=1}^{\infty} B_n \sin n\pi\eta$$

Compare each solution with the solution obtained in the example and indicate the accuracy gained.

33. A uniformly loaded rectangular plate is simply supported on three edges, clamped on the fourth edge, and point supported at x = .75a and y = .5b as shown in Fig. 7.17. Use the deflection function

$$w = \sum_{m=1}^{\infty} \sum_{n=1}^{\infty} A_{mn} \sin \frac{m\pi x}{a} \sin \frac{n\pi y}{b}$$

and the method of Ritz with Lagrange multipliers to determine an expression for the deflection of the plate.

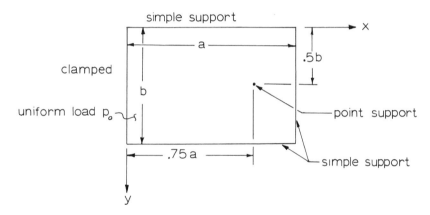

Fig. 7.17

34. A rectangular plate is simply supported on three edges and elastically supported on the fourth edge such that the deflection is zero and the moment is proportional to the slope, as shown in Fig. 7.18. Assume the lateral deflection to be of the form

$$w = (A_1 x + A_2 x^2 + A_3 x^3) \sin \frac{\pi y}{b} .$$

a. Determine the forced boundary conditions and indicate which of these are not satisfied by the assumed deflection function.

b. Explain why at least three terms of the power series ~~must be taken~~ *are desirable* in the assumed deflection function.

c. Use the assumed deflection function and the method of Ritz with Lagrange multipliers to determine an approximate expression for the lateral deflection.

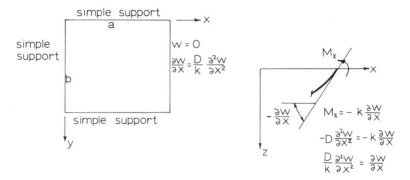

Fig. 7.18

COMPUTER ORIENTED
SOLUTIONS

Chapter 8
FINITE DIFFERENCE SOLUTIONS FOR THE LATERAL DEFLECTIONS OF PLATES

8.1 DERIVATION OF THE FINITE DIFFERENCE EQUATIONS

In the previous chapters we find that the solution techniques leading to the lateral deflections of a thin plate are often complicated, particularly if the plate is not rectangular or circular. The method of finite differences is an approximate technique which yields a simplified solution form in which the differential equations of equilibrium and the boundary conditions are replaced by a set of algebraic equations.

To describe the concept of finite differences, let's first consider the case in which we examine a function of one variable

$$F = F(x).$$

Let us assume that this function is depicted by the curve in Fig. 8.1.

Our goal is to approximate derivatives of $F(x)$ with respect to x by algebraic expressions so that we can subsequently replace any differential equation by an algebraic equation. We recall the definition of the first derivative of F with respect to x as

$$\frac{dF}{dx} = \lim_{\Delta x \to 0} \left(\frac{\Delta F}{\Delta x}\right). \tag{8.1}$$

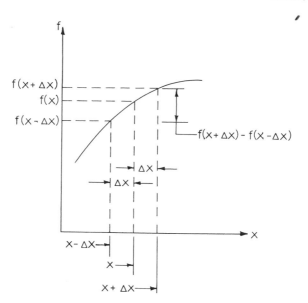

Fig. 8.1. Function of one variable

Examination of Fig. 8.1 yields the following approximation for the derivative of F with respect to x.

$$\frac{dF}{dx} \approx \frac{\Delta F}{\Delta x} = \frac{F(x + \Delta x) - F(x - \Delta x)}{2\,\Delta x} \qquad (8.2)$$

Equation (8.2) is defined as the first central finite difference approximation for dF/dx. It is apparent that this approximation becomes more accurate when the distance Δx is reduced, and in the limit as $\Delta x \to 0$ it approaches the exact expression for the first derivative. The first derivative can also be approximated by

$$\frac{dF}{dx} \approx \frac{F(x + \Delta x) - F(x)}{\Delta x} \qquad (8.3a)$$

or

$$\frac{dF}{dx} \approx \frac{F(x) - F(x - \Delta x)}{\Delta x}. \qquad (8.3b)$$

Equations (8.3a) and (8.3b) are called the first forward and first backward finite differences respectively. It is now evident that the first derivative of F with respect to x could be approximated many different ways. We could also write

$$\frac{dF}{dx} \approx \frac{F(x + 2\,\Delta x) - F(x - 2\,\Delta x)}{4\,\Delta x}. \tag{8.4}$$

We would expect Eq. (8.2) to be a better approximation than Eq. (8.4) since (8.2) is based on points closer to the point x. In this chapter we will use the central finite difference approximation given in Eq. (8.2).

Certainly only a limited number of engineering problems are described by first-order differential equations. Thus, our next step is to define an approximation for the second derivative. Let's assume that dF/dx is a continuous function as represented in Fig. 8.2.

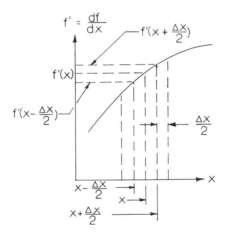

Fig. 8.2. The first derivative

Note that the divisions in Fig. 8.2 are one-half the width of the divisions in Fig. 8.1. As we proceed with the derivation, it will become evident that the divisions are reduced so that the final approximation for d^2F/dx^2 can be expressed in terms of values of F as close to the position x as possible. From Fig. 8.2 we may obtain

$$\frac{d^2F}{dx^2} = \frac{d}{dx}\left(\frac{dF}{dx}\right) \approx \frac{\Delta\left(\frac{dF}{dx}\right)}{\Delta x} \tag{8.5}$$

in which

$$\Delta\left(\frac{dF}{dx}\right)$$

is the change in dF/dx in the interval Δx. If we use the first central finite difference defined by Eq. (8.2), and, if we remember the interval is now $\Delta x/2$, we can write Eq. (8.5) in the form

$$\frac{d^2F}{dx^2} \approx \frac{\left(\dfrac{dF}{dx}\right)_{x+\Delta x/2} - \left(\dfrac{dF}{dx}\right)_{x-\Delta x/2}}{2\left(\dfrac{\Delta x}{2}\right)} \tag{8.6}$$

where

$$\left(\frac{dF}{dx}\right)_{x+\Delta x/2} \approx \frac{F(x + \Delta x) - F(x)}{\Delta x} \tag{8.7a}$$

and

$$\left(\frac{dF}{dx}\right)_{x-\Delta x/2} \approx \frac{F(x) - F(x - \Delta x)}{\Delta x}. \tag{8.7b}$$

Substitution of Eqs. (8.7a) and (8.7b) into Eq. (8.6) gives

$$\frac{d^2F}{dx^2} \approx \frac{F(x + \Delta x) - 2F(x) + F(x - \Delta x)}{(\Delta x)^2}. \tag{8.8}$$

It is now apparent that if we had not reduced the interval to $\Delta x/2$, the approximation for the second derivative would have been

$$\frac{d^2F}{dx^2} \approx \frac{1}{4(\Delta x)^2}[F(x + 2\Delta x) - 2F(x) + F(x - 2\Delta x)]. \tag{8.9}$$

This certainly is another legitimate approximation for the second derivative. However, Eq. (8.8) will generally be more accurate than Eq. (8.9), since it utilizes the function F at points nearer the point x.

Next we derive an algebraic approximation for the third derivative by obtaining the first central finite difference of the second derivative as follows.

$$\frac{d^3F}{dx^3} = \frac{d}{dx}\left(\frac{d^2F}{dx^2}\right) \approx \frac{1}{2\,\Delta x}\left\{\left[\frac{F(x + \Delta x) - 2F(x) + F(x - \Delta x)}{(\Delta x)^2}\right]_{x+\Delta x}\right.$$
$$\left. - \left[\frac{F(x + \Delta x) - 2F(x) + F(x - \Delta x)}{\Delta x^2}\right]_{x-\Delta x}\right\}$$
$$\approx \frac{1}{2\,\Delta x}[F(x + 2\,\Delta x) - 2F(x + \Delta x) + F(x)$$

$$- F(x) + 2 F(x - \Delta x) - F(x - 2 \Delta x)] \frac{1}{\Delta x^2}$$

$$\approx \frac{1}{2 \Delta x^3} [F(x + 2 \Delta x) - 2 F(x + \Delta x) + 2 F(x - \Delta x) - F(x - 2 \Delta x)] \qquad (8.10)$$

In a similar manner we develop an algebraic approximation for the fourth derivative as follows.

$$\frac{d^4F}{dx^4} = \frac{d^2}{dx^2}\left(\frac{d^2F}{dx^2}\right) \approx \frac{1}{\Delta x^2} \left\{ \left[\frac{F(x + \Delta x) - 2 F(x) + F(x - \Delta x)}{\Delta x^2} \right]_{x+\Delta x} \right.$$
$$- 2 \left[\frac{F(x + \Delta x) - 2 F(x) + F(x - \Delta x)}{\Delta x^2} \right]_{x}$$
$$\left. + \left[\frac{F(x + \Delta x) - 2 F(x) + F(x - \Delta x)}{\Delta x^2} \right]_{x-\Delta x} \right\}$$

$$\approx \frac{1}{\Delta x^4} [F(x + 2 \Delta x) - 4 F(x + \Delta x) + 6 F(x) - 4 F(x - \Delta x) + F(x - 2 \Delta x)] \qquad (8.11)$$

In the analysis of thin plates, the lateral deflections are a function of two variables:

$$w = w (x,y).$$

Thus, it is necessary to extend the concept of finite differences to a function of two variables. We refer to the two-dimensional mesh in Fig. 8.3.

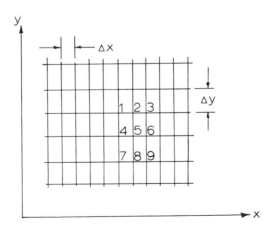

Fig. 8.3. Two-dimensional mesh

An algebraic approximation for the partial derivative of w with respect to x is obtained by taking the central finite difference of w(x,y) with respect to x and holding y constant. For the partial derivative of w with respect to y, we take the central finite difference of w(x,y) with respect to y and hold x constant. In subsequent equations we shall use equal signs exclusively, even though we know that in reality we are dealing with approximations.

$$\frac{\partial w}{\partial x} = \frac{1}{2\,\Delta x}\,[w(x + \Delta x, y) - w(x - \Delta x, y)] \qquad (8.12a)$$

$$\frac{\partial w}{\partial y} = \frac{1}{2\,\Delta y}\,[w(x, y + \Delta y) - w(x, y - \Delta y)] \qquad (8.12b)$$

If we refer to Fig. 8.3, we may write the following expressions as examples of the operations of Eqs. (8.12).

$$\left(\frac{\partial w}{\partial x}\right)_5 = \frac{1}{2\,\Delta x}\,(w_6 - w_4)$$

$$\left(\frac{\partial w}{\partial y}\right)_5 = \frac{1}{2\,\Delta y}\,(w_2 - w_8)$$

We should note that it is not necessary to have Δx equal to Δy.

Algebraic approximations of the higher order partial derivatives are presented as follows.

$$\frac{\partial^2 w}{\partial x^2} = \frac{\partial}{\partial x}\left(\frac{\partial w}{\partial x}\right) =$$
$$\frac{1}{\Delta x^2}\,[w(x + \Delta x, y) - 2\,w(x,y) + w(x - \Delta x, y)] \qquad (8.13a)$$

$$\frac{\partial^2 w}{\partial y^2} = \frac{\partial}{\partial y}\left(\frac{\partial w}{\partial y}\right) =$$
$$\frac{1}{\Delta y^2}\,[w(x, y + \Delta y) - 2\,w(x,y) + w(x, y - \Delta y)] \qquad (8.13b)$$

$$\frac{\partial^3 w}{\partial x^3} = \frac{\partial}{\partial x}\left(\frac{\partial^2 w}{\partial x^2}\right) = \frac{1}{2\,\Delta x^3}\,[w(x + 2\,\Delta x, y) - 2\,w(x + \Delta x, y)$$
$$+ 2\,w(x - \Delta x, y) - w(x - 2\,\Delta x, y)] \qquad (8.14a)$$

$$\frac{\partial^3 w}{\partial y^3} = \frac{\partial}{\partial y}\left(\frac{\partial^2 w}{\partial y^2}\right) = \frac{1}{2\,\Delta y^3}\,[w(x,y) + 2\,\Delta y) - 2\,w(x,y + \Delta y)$$
$$+ 2\,w(x,y - \Delta y) - w(x,y - 2\,\Delta y)] \qquad (8.14b)$$

$$\frac{\partial^3 w}{\partial x \partial y^2} = \frac{\partial}{\partial x}\left(\frac{\partial^2 w}{\partial x^2}\right) =$$
$$\frac{1}{2\,\Delta x \Delta y^2}\,[w(x + \Delta x, y + \Delta y) - 2\,w(x + \Delta x, y)$$
$$+ w(x + \Delta x, y - \Delta y) - w(x - \Delta x, y + \Delta y)$$
$$+ 2\,w(x - \Delta x, y) - w(x - \Delta x, y - \Delta y)] \qquad (8.15a)$$

$$\frac{\partial^3 w}{\partial x^2 \partial y} = \frac{\partial}{\partial y}\left(\frac{\partial^2 w}{\partial x^2}\right) =$$
$$\frac{1}{2\,\Delta x^2 \Delta y}\,[w(x + \Delta x, y + \Delta y) - 2\,w(x,y + \Delta y)$$
$$+ w(x - \Delta x, y + \Delta y) - w(x + \Delta x, y - \Delta y)$$
$$+ 2\,w(x,y - \Delta y) - w(x - \Delta x, y - \Delta y)] \qquad (8.15b)$$

$$\frac{\partial^4 w}{\partial x^4} = \frac{\partial^2}{\partial x^2}\left(\frac{\partial^2 w}{\partial x^2}\right) = \frac{1}{\Delta x^4}\,[w(x + 2\,\Delta x, y) - 4\,w(x + \Delta x, y)$$
$$+ 6\,w(x,y) - 4\,w(x - \Delta x, y) + w(x - 2\,\Delta x, y)] \qquad (8.16a)$$

$$\frac{\partial^4 w}{\partial y^4} = \frac{\partial^2}{\partial y^2}\left(\frac{\partial^2 w}{\partial y^2}\right) = \frac{1}{\Delta y^4}\,[w(x,y + 2\,\Delta y) - 4\,w(x,y + \Delta y)$$
$$+ 6\,w(x,y) - 4\,w(x,y - \Delta y) + w(x,y - 2\,\Delta y)] \qquad (8.16b)$$

$$\frac{\partial^4 w}{\partial x^2 \partial y^2} = \frac{\partial^2}{\partial x^2}\left(\frac{\partial^2 w}{\partial y^2}\right) =$$
$$\frac{1}{\Delta x^2 \Delta y^2}\,[w(x + \Delta x, y + \Delta y) - 2\,w(x,y + \Delta y)$$
$$+ w(x - \Delta x, y + \Delta y) - 2\,w(x + \Delta x, y) + 4\,w(x,y)$$
$$- 2\,w(x - \Delta x, y) + w(x + \Delta x, y - \Delta y)$$
$$- 2\,w(x,y - \Delta y) + w(x - \Delta x, y - \Delta y)] \qquad (8.16c)$$

Since the various partial derivatives have been expressed as finite difference approximations, now we may express the governing equation, the stress resultants, stresses, strains, and displacements as finite difference approximations. Several of these expressions are presented as follows, in which $\Delta x = \Delta y = \ell$.

$$M_x = -D\left(\frac{\partial^2 w}{\partial x^2} + \nu \frac{\partial^2 w}{\partial y^2}\right)$$

$$= -\frac{D}{\ell^2}\,[w(x+\ell,y) + w(x-\ell,y) + \nu\, w(x,y+\ell)$$

$$+ \nu\, w(x,y-\ell) - (2+2\nu)\, w(x,y)] \qquad (8.17a)$$

$$M_y = -D\left(\nu\frac{\partial^2 w}{\partial x^2} + \frac{\partial^2 w}{\partial y^2}\right)$$

$$= -\frac{D}{\ell^2}\,[\nu\, w(x+\ell,y) + \nu\, w(x-\ell,y) + w(x,y+\ell)$$

$$+ w(x,y-\ell) - (2+2\nu)\, w(x,y) \qquad (8.17b)$$

$$V_{xz} = -D\left[\frac{\partial^3 w}{\partial x^3} + (2-\nu)\frac{\partial^3 w}{\partial x \partial y^2}\right]$$

$$= -\frac{D}{2\ell^3}\,\{w(x+2\ell,y) - 2(3-\nu)\, w(x+\ell,y)$$

$$+ 2(3-\nu)\, w(x-\ell,y) - w(x-2\ell,y)$$
$$+ (2-\nu)\,[w(x+\ell,y+\ell) + w(x+\ell,y-\ell)$$
$$- w(x-\ell,y+\ell) - w(x-\ell,y-\ell)]\} \qquad (8.17c)$$

$$V_{yz} = -D\left[\frac{\partial^3 w}{\partial y^3} + (2-\nu)\frac{\partial^3 w}{\partial x^2 \partial y}\right]$$

$$= -\frac{D}{2\ell^3}\,\{w(x,y+2\ell) - 2(3-\nu)\, w(x,y+\ell)$$

$$+ 2(3-\nu)\, w(x,y-\ell) - w(x,y-2\ell)$$
$$+ (2-\nu)\,[w(x+\ell,y+\ell) + w(x-\ell,y+\ell)$$
$$- w(x+\ell,y-\ell) - w(x-\ell,y-\ell)]\} \qquad (8.17d)$$

$$\frac{p(x,y)}{D} = \frac{\partial^4 w}{\partial x^4} + 2\frac{\partial^4 w}{\partial x^2 \partial y^2} + \frac{\partial^4 w}{\partial y^4}$$

$$= \frac{1}{\ell^4}\,[20\, w(x,y) - 8w(x+\ell,y) - 8w(x-\ell,y)$$

$$- 8w(x,y+\ell) - 8w(x,y-\ell) + 2w(x+\ell,y+\ell)$$
$$+ 2w(x-\ell,y+\ell) + 2w(x+\ell,y-\ell) + 2w(x-\ell,y-\ell)$$
$$+ w(x+2\ell,y) + w(x-2\ell,y)$$
$$+ w(x,y+2\ell) + w(x,y-2\ell)] \qquad (8.18)$$

Finite difference operators are often represented in the form of patterns. Patterns for several of the operators are presented in Fig. 8.4, where the center point in each pattern of points is the point about which each operator is being written.

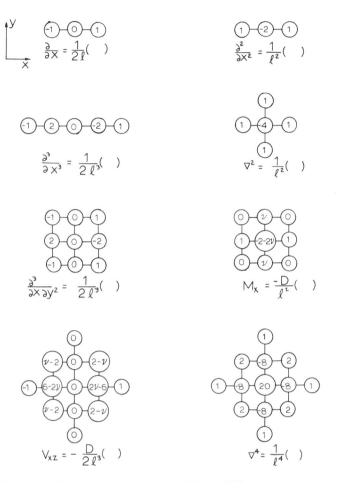

Fig. 8.4. Pattern representation of finite difference operators

8.2 SIMPLY SUPPORTED RECTANGULAR PLATES

For the special case of a simply supported plate we find that the solution is greatly simplified if we use the modified form of the equilibrium equation given by Eqs. (1.16a) and (1.16b). As a matter of convenience, we shall rewrite these equations.

$$\frac{\partial^2 w}{\partial x^2} + \frac{\partial^2 w}{\partial y^2} = \frac{M}{D} \tag{8.19a}$$

and

$$\frac{\partial^2 M}{\partial x^2} + \frac{\partial^2 M}{\partial y^2} = p \qquad (8.19b)$$

where

$$M = -\frac{M_x + M_y}{1 + \nu}$$

We recall that along any simply supported boundary of a rectangular plate we have

$$\frac{\partial^2 w}{\partial x^2} = \frac{\partial^2 w}{\partial y^2} = 0.$$

Thus, we see from Eq. (8.19a) that $M = 0$ at all points on the boundary. As a result, we can solve Eq. (8.19b) for M at all interior mesh points in the plate and then use these results to solve Eq. (8.19a) for the corresponding values of w.

As an example, let's consider the uniformly loaded simply supported plate in Fig. 8.5.

At all points on the boundary, $M = 0$. That is

$$M_1 = M_2 = M_3 = M_4 = M_5 = M_6 = M_{10} = M_{11} = M_{15}$$
$$= M_{16} = M_{20} = M_{21} = M_{22} = M_{23} = M_{24} = M_{25} = 0.$$

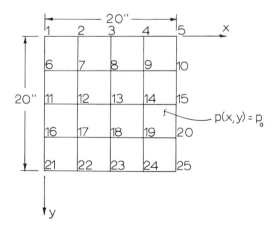

Fig. 8.5. Uniformly loaded simply supported plate

The symmetry of loading and boundary conditions leads to the following conclusions.

$$M_7 = M_9 = M_{17} = M_{19}$$

and

$$M_{12} = M_8 = M_{14} = M_{18}$$

Therefore, we may consider the only unknown values of M to be M_7, M_8, and M_{13}. We obtain the difference equations at points 7, 8, and 13 from the finite difference approximation of Eq. (8.19b) as follows.

Point 7

$$\frac{1}{(5)^2} (M_6 - 2M_7 + M_8) + \frac{1}{(5)^2} (M_2 - 2M_7 + M_{12}) = p_7$$

$$\frac{1}{25} (-4M_7 + M_8 + M_{12}) = p_7$$

$$\frac{1}{25} (-4M_7 + 2M_8) = p_0 \qquad (8.20a)$$

Point 8

$$\frac{1}{(5)^2} (M_7 - 2M_8 + M_9) + \frac{1}{(5)^2} (M_3 - 2M_8 + M_{13}) = p_8$$

$$\frac{1}{25} (M_7 - 4M_8 + M_9 + M_{13}) = p_8$$

$$\frac{1}{25} (2M_7 - 4M_8 + M_{13}) = p_0 \qquad (8.20b)$$

Point 13

$$\frac{1}{(5)^2} (M_{12} - 2M_{13} + M_{14}) + \frac{1}{(5)^2} (M_8 - 2M_{13} + M_{18}) = p_{13}$$

$$\frac{1}{25} (M_{12} - 4M_{13} + M_{14} + M_8 + M_{18}) = p_{13}$$

$$\frac{1}{25} (4M_8 - 4M_{13}) = p_0 \qquad (8.20c)$$

The simultaneous solution of Eqs. (8.20) yields

$$M_{13} = - \frac{225}{8} p_0 \qquad (8.21a)$$

$$M_7 = -\frac{275}{16} p_0 \qquad (8.21b)$$

$$M_8 = -\frac{175}{8} p_0. \qquad (8.21c)$$

We have determined the values of M at every mesh point. We shall now direct ourselves to the task of determining the values of w at each mesh point. At all points on the boundary $w = 0$; thus

$$w_1 = w_2 = w_3 = w_4 = w_5 = w_6 = w_{10} = w_{11} = w_{15}$$
$$= w_{16} = w_{20} = w_{21} = w_{22} = w_{23} = w_{24} = w_{25} = 0.$$

Again, the symmetry of loading and boundary conditions leads to the conclusions that

$$w_7 = w_9 = w_{17} = w_{19}$$

and

$$w_{12} = w_8 = w_{14} = w_{18}.$$

Therefore, we again may consider the only unknown values of w to be w_7, w_8, and w_{13}. The difference equations at points 7, 8, and 13 are obtained from the finite difference approximation of Eq. (8.19a) as follows.

Point 7

$$\frac{1}{25} (-4w_7 + 2w_8) = \frac{M_7}{D}$$

$$\frac{1}{25} (-4w_7 + 2w_8) = -\frac{275}{16D} p_0 \qquad (8.22a)$$

Point 8

$$\frac{1}{25} (2w_7 - 4w_8 + w_{13}) = \frac{M_8}{D}$$

$$\frac{1}{25} (2w_7 - 4w_8 + w_{13}) = -\frac{175}{8D} p_0 \qquad (8.22b)$$

Point 13

$$\frac{1}{25} (4w_8 - 4w_{13}) = \frac{M_{13}}{D}$$

$$\frac{1}{25} (4w_8 - 4w_{13}) = -\frac{225}{8D} p_0 \qquad (8.22c)$$

The simultaneous solution of Eqs. (8.22) yields

$$w_{13} = 644.5 \frac{p_0}{D} \qquad (8.23a)$$

$$w_7 = 341.8 \frac{p_0}{D} \qquad (8.23b)$$

$$w_8 = 468.7 \frac{p_0}{D}. \qquad (8.23c)$$

The value given by Eq. (8.23a) for the midspan deflection w_{13} is within one percent of the exact value.[1] The value given by Eq. (8.21a) for M_{13}, which is proportional to the midspan bending moment, is within five percent of the exact value.[1] If we decrease the size of the mesh increment from 5 inches to 2.5 inches, the value for M_{13} will be within one percent of the exact value.

8.3   SIMPLY SUPPORTED POLYGONAL PLATES

Boundary Conditions for a Polygonal Plate. In the previous section we determined the solution for a simply supported rectangular plate by requiring $w = 0$ and $M = 0$ at the boundary. The condition of zero deflection applies to a simply supported polygonal plate as well as to a rectangular plate. Now we will show that the condition $M = 0$ applies to the boundary of a simply supported polygonal plate as well as to a rectangular plate.

Let's consider a portion of a simply supported polygonal plate boundary as shown in Fig. 8.6.

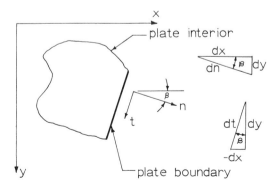

Fig. 8.6.   A portion of polygonal plate next to the boundary

The coordinate n is perpendicular to the edge of the plate, and the coordinate t is parallel to the edge. According to the chain rule of differentiation, we may write

$$\frac{\partial w}{\partial n} = \frac{\partial w}{\partial x}\frac{dx}{dn} + \frac{\partial w}{\partial y}\frac{dy}{dn} \tag{8.24a}$$

and

$$\frac{\partial w}{\partial t} = \frac{\partial w}{\partial x}\frac{dx}{dt} + \frac{\partial w}{\partial y}\frac{dy}{dt}. \tag{8.24b}$$

From Fig. 8.6, we see that

$$\frac{dx}{dn} = \cos \beta \qquad \frac{dy}{dn} = \sin \beta$$

$$\frac{dx}{dt} = -\sin \beta \qquad \frac{dy}{dt} = \cos \beta.$$

Thus, Eqs. (8.24) become

$$\frac{\partial w}{\partial n} = \frac{\partial w}{\partial x}\cos \beta + \frac{\partial w}{\partial y}\sin \beta \tag{8.25a}$$

$$\frac{\partial w}{\partial t} = -\frac{\partial w}{\partial x}\sin \beta + \frac{\partial w}{\partial y}\cos \beta. \tag{8.25b}$$

The second partial derivatives of the deflection are

$$\frac{\partial^2 w}{\partial n^2} = \frac{\partial}{\partial n}\left(\frac{\partial w}{\partial n}\right)$$

$$= \left(\cos \beta \frac{\partial}{\partial x} + \sin \beta \frac{\partial}{\partial y}\right)\left(\cos \beta \frac{\partial w}{\partial x} + \sin \beta \frac{\partial w}{\partial y}\right)$$

$$= \frac{\partial^2 w}{\partial x^2}\cos^2 \beta + 2\frac{\partial^2 w}{\partial x \partial y}\sin \beta \cos \beta + \frac{\partial^2 w}{\partial y^2}\sin^2 \beta \tag{8.26}$$

$$\frac{\partial^2 w}{\partial t^2} = \frac{\partial}{\partial t}\left(\frac{\partial w}{\partial t}\right)$$

$$= \left(-\sin \beta \frac{\partial}{\partial x} + \cos \beta \frac{\partial}{\partial y}\right)\left(-\sin \beta \frac{\partial w}{\partial x} + \cos \beta \frac{\partial w}{\partial y}\right)$$

$$= \frac{\partial^2 w}{\partial x^2}\sin^2 \beta - 2\frac{\partial^2 w}{\partial x \partial y}\sin \beta \cos \beta + \frac{\partial^2 w}{\partial y^2}\cos^2 \beta. \tag{8.27}$$

If we add Eqs. (8.26) and (8.27) we obtain

$$\frac{\partial^2 w}{\partial n^2} + \frac{\partial^2 w}{\partial t^2} = \frac{\partial^2 w}{\partial x^2} (\sin^2 \beta + \cos^2 \beta) + \frac{\partial^2 w}{\partial y^2} (\sin^2 \beta + \cos^2 \beta)$$

or

$$\frac{\partial^2 w}{\partial n^2} + \frac{\partial^2 w}{\partial t^2} = \frac{\partial^2 w}{\partial x^2} + \frac{\partial^2 w}{\partial y^2} \tag{8.28a}$$

or

$$\frac{\partial^2 w}{\partial n^2} + \frac{\partial^2 w}{\partial t^2} = \frac{M}{D}. \tag{8.28b}$$

Since the boundary is simply supported, we have

$$\frac{\partial^2 w}{\partial n^2} = \frac{\partial^2 w}{\partial t^2} = 0. \tag{8.29}$$

Thus, we see from Eq. (8.28b) that $M = 0$. We conclude that along any simply supported edge of a polygonal plate, the boundary conditions are

$$w = 0 \tag{8.30a}$$

$$M = 0. \tag{8.30b}$$

Skew Coordinates. In the analysis of thin plates, we often encounter plates with shapes that are not compatible to a rectangular mesh. An example is the skew plate in Fig. 8.7.

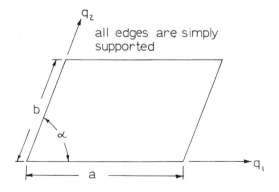

Fig. 8.7. Skew plate

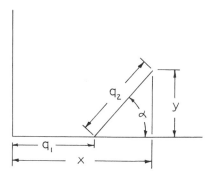

Fig. 8.8.   Coordinates x-y and q_1-q_2

It is necessary to express the governing equations

$$\nabla^2 M = p$$

and

$$\nabla^2 w = \frac{M}{D}$$

in terms of the coordinates q_1 and q_2. The transformation equations that relate the x and y coordinates to the q_1 and q_2 coordinates can be obtained with the aid of Fig. 8.8. These transformations are

$$x = q_1 + q_2 \cos \alpha$$

and

$$y = q_2 \sin \alpha$$

or

$$q_1 = x - y \frac{1}{\tan \alpha} \qquad (8.31a)$$

and

$$q_2 = y \frac{1}{\sin \alpha}. \qquad (8.31b)$$

The following expressions are obtained directly from Eqs. (8.31).

$$\frac{\partial q_1}{\partial x} = 1 \qquad (8.32a)$$

$$\frac{\partial q_1}{\partial y} = -\frac{1}{\tan \alpha} \qquad (8.32b)$$

$$\frac{\partial q_2}{\partial x} = 0 \qquad (8.32c)$$

$$\frac{\partial q_2}{\partial y} = \frac{1}{\sin \alpha} \qquad (8.32d)$$

According to the chain rule of partial differentiation we have

$$\frac{\partial w}{\partial x} = \frac{\partial w}{\partial q_1} \frac{\partial q_1}{\partial x} + \frac{\partial w}{\partial q_2} \frac{\partial q_2}{\partial x} \qquad (8.33a)$$

$$\frac{\partial w}{\partial y} = \frac{\partial w}{\partial q_1} \frac{\partial q_1}{\partial y} + \frac{\partial w}{\partial q_2} \frac{\partial q_2}{\partial y}. \qquad (8.33b)$$

Substitution of Eqs. (8.32) into (8.33) yields

$$\frac{\partial w}{\partial x} = \frac{\partial w}{\partial q_1} \qquad (8.34a)$$

and

$$\frac{\partial w}{\partial y} = -\frac{1}{\tan \alpha} \frac{\partial w}{\partial q_1} + \frac{1}{\sin \alpha} \frac{\partial w}{\partial q_2}. \qquad (8.34b)$$

The second partial derivatives of the deflection are

$$\frac{\partial^2 w}{\partial x^2} = \frac{\partial}{\partial x}\left(\frac{\partial w}{\partial x}\right) = \frac{\partial}{\partial q_1}\left(\frac{\partial w}{\partial q_1}\right) = \frac{\partial^2 w}{\partial q_1^2} \qquad (8.35a)$$

$$\begin{aligned}
\frac{\partial^2 w}{\partial y^2} &= \frac{\partial}{\partial y}\left(\frac{\partial w}{\partial y}\right) \\
&= \left(-\frac{1}{\tan \alpha}\frac{\partial}{\partial q_1} + \frac{1}{\sin \alpha}\frac{\partial}{\partial q_2}\right)\left(-\frac{1}{\tan \alpha}\frac{\partial w}{\partial q_1} + \frac{1}{\sin \alpha}\frac{\partial w}{\partial q_2}\right) \\
&= \frac{1}{\tan^2 \alpha}\frac{\partial^2 w}{\partial q_1^2} + \frac{2}{\sin \alpha \tan \alpha}\frac{\partial^2 w}{\partial q_1 \partial q_2} + \frac{1}{\sin^2 \alpha}\frac{\partial^2 w}{\partial q_2^2} \\
&= \frac{\cos^2 \alpha}{\sin^2 \alpha}\frac{\partial^2 w}{\partial q_1^2} + \frac{2\cos \alpha}{\sin^2 \alpha}\frac{\partial^2 w}{\partial q_1 \partial q_2} + \frac{1}{\sin^2 \alpha}\frac{\partial^2 w}{\partial q_2^2}. \qquad (8.35b)
\end{aligned}$$

If we add Eq. (8.35a) to (8.35b) we obtain, after simplification

$$\nabla^2 w = \frac{1}{\sin^2 \alpha}\left(\frac{\partial^2 w}{\partial q_1^2} - 2\,\frac{\partial^2 w}{\partial q_1 \partial q_2}\cos\alpha + \frac{\partial^2 w}{\partial q_2^2}\right). \qquad (8.36)$$

The finite difference equations in skew coordinates are derived with the same techniques presented in Section 8.1. Several of these equations are

$$\frac{\partial^2 w}{\partial q_1^2} = \frac{1}{\ell^2}\,[w(q_1 - \ell,q_2) - 2\,w(q_1,q_2) + w(q_1 + \ell,q_2)]$$

$$\frac{\partial^2 w}{\partial q_2^2} = \frac{1}{k^2}\,[w(q_1,q_2 - k) - 2\,w(q_1,q_2) + w(q_1,q_2 + k)]$$

$$\frac{\partial^2 w}{\partial q_1 \partial q_2} = \frac{1}{4\ell k}\,[w(q_1 - \ell,q_2 + k) - w(q_1 + \ell,q_2 + k)$$
$$- w(q_1 - \ell,q_2 - k) + w(q_1 + \ell,q_2 - k)].$$

These finite difference equations are expressed as operators in the form of patterns in Fig. 8.9. When the operators in Fig. 8.9 are applied to Eq. (8.36) we obtain the operator for ∇^2, shown in Fig. 8.10.

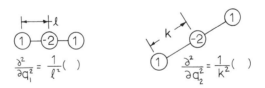

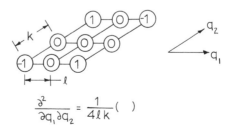

Fig. 8.9. Pattern representation of skew finite difference operators

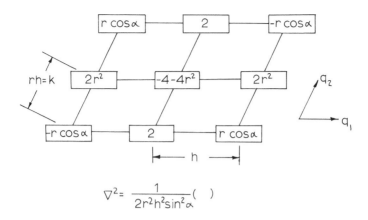

$$\nabla^2 = \frac{1}{2r^2 h^2 \sin^2 \alpha}(\quad)$$

Fig. 8.10.   Pattern representation of the ∇^2 operator in skew coordinates

Now we can simply apply the operator ∇^2 in Fig. 8.10 to the equations

$$\nabla^2 w = \frac{M}{D}$$

$$\nabla^2 M = P$$

with the conditions $M = 0$ and $w = 0$ along the boundary. The general procedure for determining the deflections is the same as described in Section 8.2.

Triangular Coordinates. Some plates have boundary configurations that are suitable for a triangular mesh as shown in Fig. 8.11. The transformation equations that relate the rectangular coordinates x–y to the triangular coordinates $q_1 - q_2 - q_3$ may be obtained with the aid of Fig. 8.11b. These transformation equations are

$$x = q_1 + q_2 \cos \alpha + q_3 \cos \beta \qquad (8.37a)$$

$$y = q_2 \sin \alpha + q_3 \sin \beta. \qquad (8.37b)$$

If we use the chain rule of partial differentiation and Eqs. (8.37), we may derive the expression for $\nabla^2 w$ in triangular coordinates using the same technique that was used in the previous section on skew coordinates.

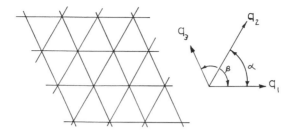

a. Triangular mesh

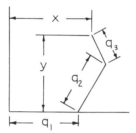

b. Triangular coordinates

Fig. 8.11.   Triangular mesh and coordinates $q_1 - q_2 - q_3$

$$\nabla^2 w = \frac{2}{3}\left(\frac{\partial^2 w}{\partial q_1^2} + \frac{\partial^2 w}{\partial q_2^2} + \frac{\partial^2 w}{\partial q_3^2}\right) \tag{8.38}$$

For the case of equilateral triangular coordinates, the finite difference operator for ∇^2 is given in Fig. 8.12. Again, we can apply the ∇^2 operator in Fig. 8.12 to the equations

$$\nabla^2 w = \frac{M}{D}$$

$$\nabla^2 M = p$$

with the conditions $M = 0$ and $w = 0$ along the boundary, and the procedure for determining the deflections is the same as described in Section 8.2.

Rectangular Coordinates. In the analysis of a polygonal plate in which a rectangular mesh is used, the mesh points may not coincide with the plate boundary, as illustrated in Fig. 8.13. The previously de-

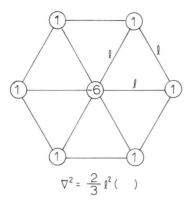

$$\nabla^2 = \frac{2}{3}\ell^2 (\quad)$$

Fig. 8.12. Pattern representation of the ∇^2 operator in equilateral triangular coordinates

fined finite difference approximation for the ∇^2 operator, given in Fig. 8.4d, does not apply to point "0" in Fig. 8.13 since ℓ_1 and ℓ_2 are not equal to ℓ. To develop the appropriate finite difference approximation of $\nabla^2 w$, we assume the expression for w in the region immediately surrounding point "0" to be

$$w(x,y) \;=\; a_0 + a_1 x + a_2 y + a_3 x^2 + a_4 y^2. \qquad (8.39)$$

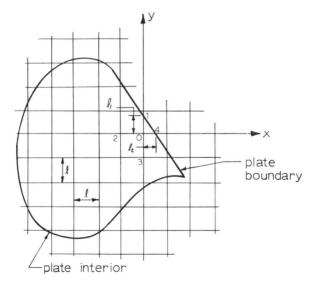

Fig. 8.13. Polygonal plate section with rectangular mesh

The origin of the coordinate system corresponding to Eq. (8.39) is at point "0." The five coefficients a_0, a_1, a_2, a_3, and a_4 may be evaluated from the following five conditions.

$$w \Big|_{\substack{x = 0 \\ y = 0}} = w_0 \tag{8.40a}$$

$$w \Big|_{\substack{x = 0 \\ y = \ell_1}} = w_1 \tag{8.40b}$$

$$w \Big|_{\substack{x = -\ell \\ y = 0}} = w_2 \tag{8.40c}$$

$$w \Big|_{\substack{x = 0 \\ y = -\ell}} = w_3 \tag{8.40d}$$

$$w \Big|_{\substack{x = \ell_2 \\ y = 0}} = w_4 \tag{8.40e}$$

If we substitute the conditions given by Eqs. (8.40) into (8.39) we obtain

$$w_0 = a_0 \tag{8.41a}$$

$$w_1 = a_0 + a_2\ell_1 + a_4\ell_1^2 \tag{8.41b}$$

$$w_2 = a_0 - a_1\ell + a_3\ell^2 \tag{8.41c}$$

$$w_3 = a_0 - a_2\ell + a_4\ell^2 \tag{8.41d}$$

$$w_4 = a_0 + a_1\ell_2 + a_3\ell_2^2. \tag{8.41e}$$

The simultaneous solution of Eqs. (8.41) yields

$$a_0 = w_0 \tag{8.42a}$$

$$a_1 = \frac{\ell^2 (w_4 - w_0) + \ell_2^2 (w_0 - w_2)}{\ell \ell_2 (\ell + \ell_2)} \tag{8.42b}$$

$$a_2 = \frac{\ell^2 (w_1 - w_0) + \ell_1^2 (w_0 - w_3)}{\ell \ell_1 (\ell + \ell_1)} \tag{8.42c}$$

$$a_3 = \frac{\ell(w_4 - w_0) - \ell_2(w_0 - w_2)}{\ell\ell_2(\ell + \ell_2)}$$ (8.42d)

$$a_4 = \frac{\ell(w_1 - w_0) - \ell_1(w_0 - w_3)}{\ell\ell_1(\ell + \ell_1)}.$$ (8.42e)

If we substitute the expression for w given by Eq. (8.39) into the equation $\nabla^2 w = M/D$ we obtain

$$2\,a_3 + 2\,a_4 = \frac{M}{D}$$

or

$$2\left[\frac{\ell(w_4 - w_0) - \ell_2(w_0 - w_2)}{\ell\ell_2(\ell + \ell_2)}\right.$$
$$\left. + \frac{\ell(w_1 - w_0) - \ell_1(w_0 - w_3)}{\ell_1\ell_2(\ell_1 + \ell_2)}\right] = \frac{M_0}{D}.$$ (8.43)

If we replace w by M and M_0/D by p_0 in Eq. (8.43), we obtain the finite difference approximation of the expression $\nabla^2 M = p$:

$$2\left[\frac{\ell(M_4 - M_0) - \ell_2(M_0 - M_2)}{\ell\ell_2(\ell + \ell_2)}\right.$$
$$\left. + \frac{\ell(M_1 - M_0) - \ell_1(M_0 - M_3)}{\ell_1\ell_2(\ell_1 + |\ell_2)}\right] = -p_0.$$ (8.44)

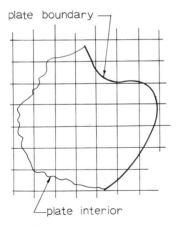

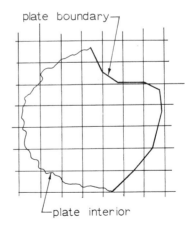

| a. Curved boundary | b. Straight line approximation of boundary |

Fig. 8.14.   Curved boundary and its straight line approximation

If we use Eq. (8.43) and the boundary conditions $M = 0$ and $w = 0$ along the boundary, we may analyze any simply supported polygonal plate with a rectangular mesh according to the procedure described in Section 8.2.

If the plate boundary is curved, a good approximate solution can be obtained by using a rectangular mesh with straight line approximations of the boundary as shown in Fig. 8.14.

8.4 RECTANGULAR PLATES WITH ARBITRARY BOUNDARY CONDITIONS

In this section we shall address ourselves to the method of finite differences as applied to the lateral deflections of rectangular plates with arbitrary boundary conditions. Our solution technique requires that equilibrium be satisfied at all points and that the boundary conditions be satisfied. Let's begin by examining the boundary conditions. We must determine the finite difference approximations for the various boundary conditions. Three types of boundary conditions are considered: (1) simply supported, (2) clamped, and (3) free.

Simply Supported Edge Conditions. For a simply supported edge parallel to the y axis, the finite difference approximations of the boundary conditions $w = 0$ and $M_x = 0$ or $\partial^2 w / \partial x^2 = 0$ are

$$w(x,y) = 0 \qquad (8.45a)$$

and

$$w(x + \Delta x, y) - 2w(x,y) + w(x - \Delta x, y) = 0$$

or

$$w(x + \Delta x, y) = - w(x - \Delta x, y). \qquad (8.45b)$$

We see that the application of Eq. (8.45b) to points along the boundary requires additional mesh points that fall outside of the plate boundary. The meaning of the finite difference approximation given by Eq. (8.45b) becomes clear if we recall that $\partial^2 w / \partial x^2 = 0$ implies that the curvature of the plate with respect to the x axis is zero. That is, the plate becomes flat at the edge as illustrated in Fig. 8.15. Thus, we can see that

$$w(x - \Delta x, y) = - w(x + \Delta x, y)$$

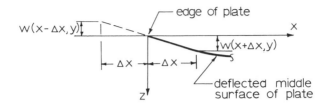

Fig. 8.15.   Model of simply supported boundary condition

is a good approximation for $\partial^2 w / \partial x^2 = 0$, providing Δx is small enough.

Equations (8.45a) and (8.45b) must be applied to each point on the simply supported edge. As an example, let us assume that the boundary of the plate in Fig. 8.16 is simply supported. Equations (8.45a) and (8.45b) applied to each point on the simply supported boundary yield

$$w_2 = w_6 = w_{10} = w_{14} = w_{18} = 0$$

$$w_1 = - w_3$$

$$w_5 = - w_7$$

$$w_9 = - w_{11}$$

$$w_{13} = - w_{15}$$

$$w_{17} = - w_{19}.$$

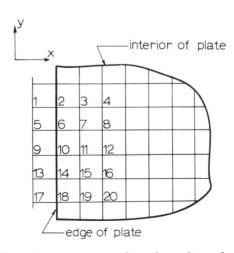

Fig. 8.16.   Rectangular mesh at plate edge or boundary

For all points on a simply supported edge parallel to the x axis, the finite difference approximations of the boundary conditions $w = 0$ and $M_y = 0$ or $\partial^2 w / \partial y^2 = 0$ are

$$w(x,y) = 0 \qquad (8.45c)$$

and

$$w(x, y + \Delta y) - 2w(x,y) + w(x, y - \Delta y) = 0$$

or

$$w(x, y + \Delta y) = -w(x, y - \Delta y). \qquad (8.45d)$$

Clamped Edge Conditions. For a clamped edge parallel to the y axis, the finite difference approximations of the boundary conditions $w = 0$ and $\partial w / \partial x = 0$ are

$$w(x,y) = 0 \qquad (8.46a)$$

and

$$w(x + \Delta x, y) - w(x - \Delta x, y) = 0$$

or

$$w(x + \Delta x, y) = w(x - \Delta x, y). \qquad (8.46b)$$

Again we find that mesh points outside the plate boundary are required according to Eq. (8.46b). The physical interpretation of Eq. (8.46b) is illustrated in Fig. 8.17.

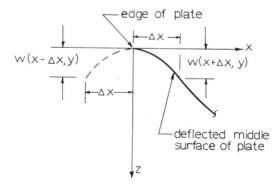

Fig. 8.17. Model of clamped boundary condition

Equations (8.46a) and (8.46b) must be applied to each point on the clamped boundary. As an example, let us assume that the boundary of the plate in Fig. 8.16 is clamped. Equations (8.46a) and (8.46b) applied to each point on the clamped boundary yield

$$w_2 = w_6 = w_{10} = w_{14} = w_{18} = 0$$

$$w_1 = w_3$$

$$w_5 = w_7$$

$$w_9 = w_{11}$$

$$w_{13} = w_{15}$$

$$w_{17} = w_{19}.$$

If the edge is parallel to the x axis, the finite difference approximations of the boundary conditions become

$$w(x,y) = 0 \tag{8.46c}$$

and

$$w(x,y + \Delta y) = w(x,y - \Delta y). \tag{8.46d}$$

Free Edge Conditions. For a free edge parallel to the y axis, the boundary conditions are $M_x = V_{xz} = 0$. The finite difference approximation of $M_x = 0$, obtained from Eq. (8.17a), is

$$w(x + \Delta x,y) + w(x - \Delta x,y) + \nu w(x,y + \Delta y)$$
$$+ \nu w(x,y - \Delta y) - (2 + 2\nu) w(x,y) = 0. \tag{8.47a}$$

The operation expressed by Eq. (8.47a) may be exhibited by the pattern in Fig. 8.18 where the center point in the pattern is the point to

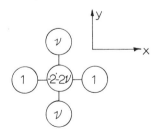

Fig. 8.18.   Finite difference pattern for M_x

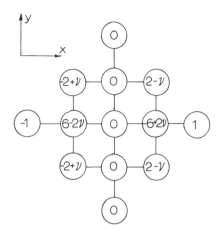

Fig. 8.19. Finite difference pattern for V_{xz}

which the operation is applied. The finite difference approximation of $V_{xz} = 0$, obtained from Eq. (8.17c), is

$$w(x + 2\Delta x, y) - 2(3 - \nu) w(x + \Delta x, y) + 2(3 - \nu) w(x - \Delta x, y)$$
$$- w(x - 2\Delta x, y) + (2 - \nu) w(x + \Delta x, y + \Delta y)$$
$$+ (2 - \nu) w(x + \Delta x, y - \Delta y) - (2 - \nu) w(x - \Delta x, y + \Delta y)$$
$$- (2 - \nu) w(x - \Delta x, y - \Delta y) = 0. \tag{8.47b}$$

The operation expressed by Eq. (8.47b) may be exhibited by the pattern shown in Fig. 8.19 where, again, the center point in the pattern is the point to which the operation is applied. Equations (8.47a) and (8.47b) must be applied to each point on a free boundary. Again we find that mesh points outside of the plate boundary are required according to Eqs. (8.47a) and (8.47b). Equivalent expressions for the finite difference approximations of the boundary conditions on a free edge parallel to the x axis can be obtained from Eqs. (8.17b) and (8.17d).

Example

We shall direct ourselves to the problem of determining the lateral deflections of the rectangular plate in Fig. 8.20.

The finite difference approximation of the governing differential equation of equilibrium, given by Eq. (8.18), must be applied to all points on the plate except for those on the simply supported and clamped boundaries. At those points on the simply supported and

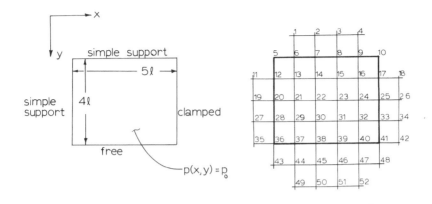

a. Rectangular mesh   b. Oblique mesh

Fig. 8.20. Rectangular plate with uniform lateral load p(x,y) = p_0

clamped boundaries, equilibrium is satisfied by the condition that w =
0. Thus, Eq. (8.18) must be applied to all interior points and also to the
Points 37, 38, 39, and 40 on the free edge. The finite difference equations obtained by applying Eq. (8.18) to all interior points and to the
Points 37, 38, 39, and 40 on the free edge involve the fifty-two points
illustrated in the mesh of Fig. 8.16. Thus, we must obtain a total of
fifty-two independent equations in terms of these fifty-two values of the
deflection, as follows.

Governing Equation of Equilibrium ($\nabla^4 w = p/D$) . . . 16 Equations

$$20w_{13} + w_1 + w_{29} + w_{11} + w_{15} + 2\left(w_5 + w_7 + w_{20} + w_{22}\right)$$
$$- 8\left(w_6 + w_{21} + w_{12} + w_{14}\right) = \frac{p_0 \ell^4}{D}$$

$$20w_{14} + w_2 + w_{30} + w_{12} + w_{16} + 2\left(w_6 + w_8 + w_{21} + w_{23}\right)$$
$$- 8\left(w_7 + w_{23} + w_{13} + w_{15}\right) = \frac{p_0 \ell^4}{D}$$

$$20w_{15} + w_3 + w_{31} + w_{13} + w_{17} + 2\left(w_7 + w_9 + w_{22} + w_{24}\right)$$
$$- 8\left(w_8 + w_{23} + w_{14} + w_{16}\right) = \frac{p_0 \ell^4}{D}$$

$$20w_{16} + w_4 + w_{32} + w_{14} + w_{18} + 2\left(w_8 + w_{10} + w_{23} + w_{25}\right)$$
$$- 8\left(w_9 + w_{15} + w_{17} + w_{24}\right) = \frac{p_0 \ell^4}{D}$$

$$20w_{21} + w_6 + w_{37} + w_{19} + w_{23} + 2\,(w_{12} + w_{14} + w_{28} + w_{30})$$
$$- 8\,(w_{13} + w_{20} + w_{29} + w_{22}) = \frac{p_0 \ell^4}{D}$$

$$20w_{22} + w_7 + w_{38} + w_{20} + w_{24} + 2\,(w_{13} + w_{15} + w_{29} + w_{31})$$
$$- 8\,(w_{14} + w_{21} + w_{30} + w_{23}) = \frac{p_0 \ell^4}{D}$$

$$20w_{23} + w_8 + w_{39} + w_{21} + w_{25} + 2\,(w_{14} + w_{16} + w_{30} + w_{32})$$
$$- 8\,(w_{15} + w_{22} + w_{31} + w_{24}) = \frac{p_0 \ell^4}{D}$$

$$20w_{24} + w_9 + w_{40} + w_{22} + w_{26} + 2\,(w_{15} + w_{17} + w_{31} + w_{33})$$
$$- 8\,(w_{16} + w_{23} + w_{25} + w_{32}) = \frac{p_0 \ell^4}{D}$$

$$20w_{29} + w_{13} + w_{44} + w_{27} + w_{31} + 2\,(w_{20} + w_{22} + w_{36} + w_{38})$$
$$- 8\,(w_{21} + w_{28} + w_{30} + w_{37}) = \frac{p_0 \ell^4}{D}$$

$$20w_{30} + w_{14} + w_{45} + w_{28} + w_{32} + 2\,(w_{21} + w_{23} + w_{37} + w_{39})$$
$$- 8\,(w_{22} + w_{29} + w_{31} + w_{38}) = \frac{p_0 \ell^4}{D}$$

$$20w_{31} + w_{15} + w_{46} + w_{29} + w_{33} + 2\,(w_{22} + w_{24} + w_{38} + w_{40})$$
$$- 8\,(w_{23} + w_{30} + w_{32} + w_{39}) = \frac{p_0 \ell^4}{D}$$

$$20w_{32} + w_{16} + w_{47} + w_{30} + w_{34} + 2\,(w_{23} + w_{25} + w_{39} + w_{41})$$
$$- 8\,(w_{24} + w_{31} + w_{33} + w_{40}) = \frac{p_0 \ell^4}{D}$$

$$20w_{37} + w_{21} + w_{49} + w_{35} + w_{39} + 2\,(w_{28} + w_{30} + w_{43} + w_{45})$$
$$- 8\,(w_{29} + w_{36} + w_{38} + w_{44}) = \frac{p_0 \ell^4}{D}$$

$$20w_{38} + w_{22} + w_{50} + w_{36} + w_{40} + 2\,(w_{29} + w_{31} + w_{44} + w_{46})$$
$$- 8\,(w_{30} + w_{37} + w_{39} + w_{45}) = \frac{p_0 \ell^4}{D}$$

$$20w_{39} + w_{23} + w_{51} + w_{37} + w_{41} + 2\,(w_{30} + w_{32} + w_{45} + w_{47})$$
$$- 8\,(w_{31} + w_{38} + w_{40} + w_{46}) = \frac{p_0 \ell^4}{D}$$

$$20w_{40} + w_{24} + w_{52} + w_{38} + w_{42} + 2\left(w_{31} + w_{33} + w_{46} + w_{48}\right)$$
$$- 8\left(w_{32} + w_{39} + w_{41} + w_{47}\right) = \frac{p_0 \ell^4}{D}$$

Boundary Condition of Zero Deflection (w = 0) on the Simply Supported and Clamped Edges . . . 14 Equations

$$w_{36} = w_{28} = w_{20} = w_{12} = w_5 = w_6 = w_7$$
$$= w_8 = w_9 = w_{10} = w_{17} = w_{25} = w_{33} = w_{41} = 0$$

Boundary Condition of Zero Moment ($\partial^2 w/\partial x^2 = 0$) on the Simply Supported Edge Parallel to y Axis . . . 4 Equations

$$w_{35} = - w_{37}$$

$$w_{27} = - w_{29}$$

$$w_{19} = - w_{21}$$

$$w_{11} = - w_{13}$$

Boundary Condition of Zero Moment ($\partial^2 w/\partial y^2 = 0$) on the Simply Supported Edge Parallel to x Axis . . . 4 Equations

$$w_1 = - w_{13}$$

$$w_2 = - w_{14}$$

$$w_3 = - w_{15}$$

$$w_4 = - w_{16}$$

Boundary Condition of Zero Slope ($\partial w/\partial x = 0$) on the Clamped Edge . . . 4 Equations

$$w_{18} = w_{16}$$

$$w_{24} = w_{26}$$

$$w_{22} = w_{34}$$

$$w_{40} = w_{42}$$

Boundary Condition of Zero Shear ($V_{yz} = 0$) on Free Edge at Points 37, 38, 39, 40 . . . 4 Equations

$$w_{49} - w_{21} + (\nu - 2)\,(w_{28} + w_{30} - w_{43} - w_{45})$$
$$+ (6 - 2\nu)\,(w_{29} - w_{44}) = 0$$

$$w_{50} - w_{22} + (\nu - 2)\,(w_{29} + w_{31} - w_{44} - w_{46})$$
$$+ (6 - 2\nu)\,(w_{30} - w_{45}) = 0$$

$$w_{51} - w_{23} + (\nu - 2)\,(w_{30} + w_{32} - w_{45} - w_{47})$$
$$+ (6 - 2\nu)\,(w_{31} - w_{46}) = 0$$

$$w_{52} - w_{24} + (\nu - 2)\,(w_{31} + w_{33} - w_{46} - w_{48})$$
$$+ (6 - 2\nu)\,(w_{32} - w_{47}) = 0$$

Boundary Condition of Zero Moment ($M_y = 0$) on Free Edge at Points 37, 38, 39, 40 . . . 4 Equations

$$- 2(1 + \nu)\,w_{37} + w_{29} + w_{44} + \nu(w_{36} + w_{38}) = 0$$

$$- 2(1 + \nu)\,w_{38} + w_{30} + w_{45} + \nu(w_{37} + w_{39}) = 0$$

$$- 2(1 + \nu)\,w_{39} + w_{31} + w_{46} + \nu(w_{38} + w_{40}) = 0$$

$$- 2(1 + \nu)\,w_{40} + w_{32} + w_{47} + \nu(w_{39} + w_{41}) = 0$$

Continuity of Zero Curvature . . . 2 Equations

Up to this point we have a total of fifty independent equations in terms of the fifty-two unknown values of w. We know that the plate is flat along the edges between Points 5 and 36 and between Points 10 and 41. Thus, we have the two expressions $w_{43} = - w_{28}$ and $w_{48} = - w_{33}$ in accordance with Fig. 8.15. Since $w_{33} = w_{28} = 0$, these two expressions reduce to

$$w_{48} = 0$$

$$w_{43} = 0.$$

We now have a total of fifty-two independent algebraic equations from which we can determine the values of the deflection at the fifty-two mesh points. Many of the equations in this set of fifty-two are quite elementary, such as those resulting from the boundary conditions at the clamped and simply supported edges. Thus, the set may easily be reduced to twenty-four independent equations in twenty-four unknowns.

There are many methods to solve systems of algebraic equations simultaneously. These methods may be divided into two major categories: (1) direct methods, and (2) iteration methods. The direct methods such as the Gauss reduction method,[30] the Gauss-Jordon reduction method,[30] and the Cholesky method[29] are direct in the sense that they lead to a correct answer at the end of a particular number of steps, this number being a function of the method used. One of the disadvantages of the direct methods is that errors are accumulative; thus, these methods are not practical for hand computation where a large number of equations are involved. The iteration methods, such as the Gauss-Seidel method[29] and the method of relaxation are self-correcting methods. If an error is made, these methods will still converge to the correct solution; thus, they are practical for hand computation providing the number of equations is not too large. Both the direct methods and the iteration methods are conveniently adaptable to digital computer solutions.

Problems

35. The plate in Fig. 8.21 is simply supported on all edges and is under the action of a uniform load p_0. Determine the deflections at the mesh points if

    a. the mesh is rectangular,

    b. the mesh is oblique.

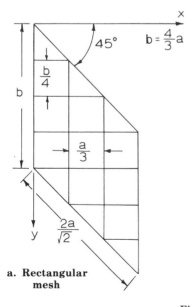

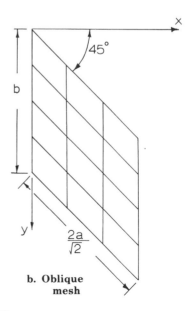

a. Rectangular mesh

b. Oblique mesh

Fig. 8.21

36. Consider the triangular plate in Fig. 8.22. If h = .2, what must be the value of a uniform load p_0 which results in an .08 in deflection at Point "one?"

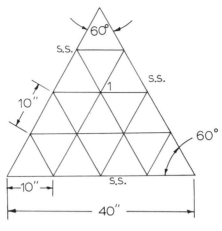

Fig. 8.22

37. Let's use the rectangular mesh shown for the circular plate in Fig. 8.23. We are given

$$\text{diameter} = 32 \text{ in}$$
$$w_{max} = h/2$$
$$p = 100 \ (r - 16) \ \text{lb/in}^2$$

in which

      r = radial distance from the center of the plate. What is the necessary thickness, h?

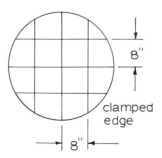

Fig. 8.23

38. Write all the appropriate algebraic equations which will lead to the deflections at the mesh points of the plate in Fig. 8.24.

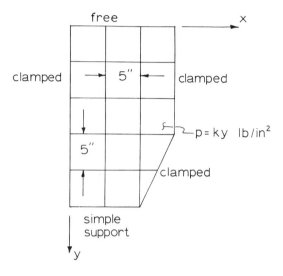

Fig. 8.24

Chapter 9
FINITE ELEMENT SOLUTIONS FOR THE LATERAL DEFLECTIONS OF PLATES

9.1 INTRODUCTION

In this chapter we shall consider the application of the finite element method to the analysis of thin plates. This method is based on approximating the behavior of a continuum by the behavior of a discrete element representation of the continuum; that is, the entire plate is represented with a model that consists of a finite number of idealized plate elements interconnected at discrete points called nodes. The analysis becomes more accurate as a larger number of elements are taken in the model representing the continuum. A basic understanding of the finite element method is presumed by the authors; thus, our discussion dwells not on the general method but rather on the stiffness or bending properties of finite plate elements. These stiffness properties of a plate element are represented in the form of a matrix which relates lateral deflections and slopes at the interconnecting points or nodes to the corresponding forces and moments acting on the element at the nodes. The reader who may need to review the general finite element method of analysis is referred to texts by Przemieniecki,[32] and Zienkiewicz and Cheung.[33]

The finite element method of analysis has been in use for approximately two decades, and has been widely accepted in the field of structural mechanics. Since this method is highly dependent upon the digital computer for implementing the necessary matrix operations, its development has closely paralleled that of digital computer systems. For every finite plate element there is a relationship between the deflections and slopes at the nodal (interconnecting) points and the corresponding forces and moments at the nodes. This relationship may be presented in the form of a matrix called a stiffness matrix. In this chapter, we shall first discuss the stiffness matrices for both rectangular and triangular plate elements. Then we shall present the results of several numerical examples of the finite element method of plate analysis. Next, we shall present stiffness matrices for plate elements with anisotropic properties. Finally, we shall take up the problems of large deflections of plates by the finite element method.

9.2   RECTANGULAR PLATE ELEMENTS

Let us consider the flat rectangular plate in Fig. 9.1 which is subdivided into nine rectangular finite elements. The geometrical position of a rectangular plate element is determined by the four nodal points (i,j,k,l) and the straight line boundaries. We assume that each nodal point of an element has a transverse deflection w, a rotation θ_x about an x axis, and a rotation θ_y about a y axis. This deflection and the two rotations are referred to as a set of nodal displacements. We also assume that a transverse force $\overline{F}_w$, a moment $\overline{M}_x$ about an x axis, and a moment $\overline{M}_y$ about a y axis act at each nodal point of an element. Do not confuse

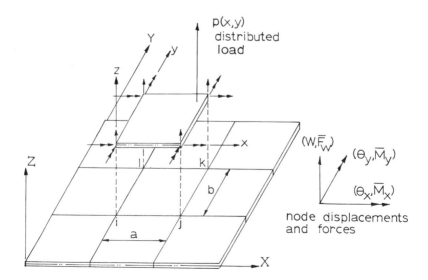

Fig. 9.1.   Rectangular plate elements

the nodal moments $\overline{M}_x$ and $\overline{M}_y$ with the stress resultants M_x and M_y developed in Chapter 1. This nodal force and the two nodal moments are referred to as a set of nodal forces. The nodal displacements and nodal forces are assumed to be positive in the directions shown in Fig. 9.1. The coordinate set (X, Y, Z) is generally referred to as the global set for the complete structure, while the set (x, y, z) is called the local set for the finite element under consideration.

In this section we develop the relationship between the four sets of nodal displacements and the four sets of nodal forces for a rectangular plate element. This relationship is presented in the form of a matrix which we refer to as an element stiffness matrix.

We begin by selecting the following polynomial expression for the deflection of the element.

$$w = a_1 + a_2x + a_3y + a_4x^2 + a_5xy + a_6y^2 + a_7x^3$$
$$+ a_8x^2y + a_9xy^2 + a_{10}y^3 + a_{11}x^3y + a_{12}xy^3 \qquad (9.1)$$

where the quantities $a_1, a_2, \ldots a_{12}$ are unknown parameters. We introduce the brackets [ ] to represent a matrix, the braces $\{\ \}$ to represent a matrix of one column, and $[\ ]^T$ or $\{\ \}^T$ to represent the transpose of a matrix. The deflection function may be symbolically expressed in matrix form as

$$w = \{P\}^T \{a\} \qquad (9.2a)$$

where

$$\{P\} = \left\{ \begin{array}{c} 1 \\ x \\ y \\ x^2 \\ . \\ . \\ . \\ x^3y \\ xy^3 \end{array} \right\} \qquad (9.2b)$$

$$\{a\} = \left\{ \begin{array}{c} a_1 \\ a_2 \\ . \\ . \\ . \\ a_{12} \end{array} \right\} . \qquad (9.2c)$$

This deflection function must possess certain basic properties. The function must contain at least as many unknown parameters as the number of nodal displacements (degrees of freedom) for the rectangular element. There are three degrees of freedom (w, θ_x, θ_y) at each of the four nodal points (i,j,k,l); thus, the function must have at least twelve unknown parameters. If the deflection function for the rectangular plate contains more than twelve unknown parameters, the derivation of the stiffness matrix involves the solution of a set of overdefined linear algebraic equations in these parameters, which we avoid in this presentation.

The function must contain a constant term, a term linear in x, and a term linear in y to represent the rigid body displacement of the element. The total displacement of the element from its initial position to its final position can be thought of as including a rigid body displacement to some intermediate position, plus an elastic displacement with respect to this intermediate position. The rigid body displacement consists of three components: (1) a translation in the z direction, (2) a rotation about an x axis, and (3) a rotation about a y axis. These components are accounted for by the first three terms in Eq. (9.1). Note that these three terms give zero strain according to Eqs. (1.7).

The situation is sometimes encountered where the solution to a problem is formed such that the set of local coordinates for a particular element has a different geometric orientation than the set of local coordinates for another element. That is, the x and y coordinates of one element may be rotationally displaced from those of another element by an arbitrary angle. Recall that stress, as well as many other quantities, is a function of the lateral deflections. Thus, if the general expression for stress is to be the same for all elements, then the general form of the selected deflection function must be invariant to any rotational transformation of coordinates. The general form of the deflection function given by Eq. (9.1) is invariant to 90 degree (or a multiple thereof) rotational transformations.

Before we discuss the next required property of the selected deflection function, we form the matrix expression for the parameters a_1, a_2, . . . , a_{12}. First, we write the general expression for a set of nodal displacements.

$$\{u_m\} = \begin{Bmatrix} w \\ \theta_x \\ \theta_y \end{Bmatrix}_{node\ m} = \begin{Bmatrix} w(x_m, y_m) \\ \dfrac{\partial}{\partial y} w(x_m, y_m) \\ -\dfrac{\partial}{\partial x} w(x_m, y_m) \end{Bmatrix}, \quad m = i,j,k,l \qquad (9.3)$$

Next, we combine the expressions for each of the four sets of nodal displacements of Eq. (9.3) into one matrix equation as

$$\{u^n\} = [C]\{a\} \qquad (9.4a)$$

where

$$\{u^n\} = \begin{Bmatrix} u_i \\ u_j \\ u_k \\ u_l \end{Bmatrix}. \qquad (9.4b)$$

If we solve Eq. (9.4a) for $\{a\}$, we obtain

$$\{a\} = [C^{-1}]\{u^n\} \qquad (9.5a)$$

where the matrix $[C^{-1}]$ is the inverse of the matrix $[C]$. We see that Eq. (9.5a) requires a matrix inverse operation; thus, the deflection function must be chosen so that the square matrix $[C]$ of dimension (12,12) is non-singular. For the case of a rectangular finite element with the coordinate system displayed in Fig. 9.1 and having dimensions of a and b in the x and y directions respectively, the matrix $[C^{-1}]$ has the form given by Eq. (9.5b).

$$[C^{-1}] = \frac{1}{a^3b^3}
\begin{bmatrix}
a^3b^3 & 0 & 0 & 0 & 0 & 0 & 0 & 0 & 0 & 0 & 0 & 0 \\
0 & 0 & -a^3b^3 & 0 & 0 & 0 & 0 & 0 & 0 & 0 & 0 & 0 \\
0 & a^3b^3 & 0 & 0 & 0 & 0 & 0 & 0 & 0 & 0 & 0 & 0 \\
-3ab^3 & 0 & 2a^2b^3 & 3ab^3 & 0 & a^2b^3 & 0 & 0 & 0 & 0 & 0 & 0 \\
-a^2b^2 & -a^2b^3 & a^2b^2 & a^2b^2 & a^2b^3 & 0 & -a^2b^2 & 0 & 0 & a^2b^2 & 0 & -a^3b^2 \\
-3a^3b & -2a^3b^2 & 0 & 0 & 0 & 0 & 0 & 0 & 0 & 3a^3b & -a^3b^2 & 0 \\
2b^3 & 0 & -ab^3 & -2b^3 & 0 & -ab^3 & 0 & 0 & 0 & 0 & 0 & 0 \\
3ab^2 & 0 & -2a^2b^2 & -3ab^2 & 0 & -a^2b^2 & 3ab^2 & 0 & a^2b^2 & -3ab^2 & 0 & 2a^2b^2 \\
3a^2b & 2a^2b^2 & 0 & -3a^2b & -2a^2b^2 & 0 & 3a^2b & -a^2b^2 & 0 & -3a^2b & a^2b^2 & 0 \\
2a^3 & a^3b & 0 & 0 & 0 & 0 & 0 & 0 & 0 & -2a^3 & a^3b & 0 \\
-2b^2 & 0 & ab^2 & 2b^2 & 0 & ab^2 & -2b^2 & 0 & -ab^2 & 2b^2 & 0 & -ab^2 \\
-2a^2 & -a^2b & 0 & 2a^2 & a^2b & 0 & -2a^2 & a^2b & 0 & 2a^2 & -a^2b & 0
\end{bmatrix}
\qquad (9.5b)$$

To complete our discussion of the properties of the deflection function, consider a particular nodal point where four adjacent elements are joined. In the process of formulating the solution to a problem, we require the deflections of each of the four elements at that point to be equal. We make similar requirements for rotations (slopes). These requirements are referred to as compatibility requirements of nodal displacements. A desirable property of the selected deflection function involves the compatibility of deflections and rotations along the common boundary line of adjacent elements. As an example, let us consider the general expression for deflection along a particular boundary line parallel to the y axis. We find that the expression for the deflection of each adjacent element along this common boundary is a cubic in the variable y. A cubic is uniquely defined by four constants as follows.

$$w = A_1 + A_2y + A_3y^2 + A_4y^3$$

The values of deflection and the normal rotation θ_x at each of the two nodal points at the extremities of the element boundary provide the necessary information to determine these constants. Since this information is identical for the adjacent elements, the deflection functions of the adjacent elements along the common boundary are compatible (equal). It therefore follows that the normal rotations of adjacent elements along the common boundary are also compatible. The general expression for the tangential rotation θ_y of each adjacent element along this common boundary is also a cubic in the variable y. This cubic is defined by four different constants. The value of the tangential rotation θ_y at each of the two nodal points at the extremities of the element boundary does not provide sufficient information to determine these four constants. Thus, we conclude that tangential rotations of the adjacent elements along the common boundary are not compatible for our selected deflection function of Eq. (9.1). However, experience has indicated that this deflection function gives highly satisfactory results for a wide set of problems. Compatibility of tangential rotations is very difficult to establish. Various investigators[33] have studied the use of corrective functions to circumvent this difficulty; however, these functions have shown little consistent value for providing increased accuracy or computational efficiency. Irons and Draper[34] have advanced the argument that plate bending elements should not be connected at the nodes, but rather along the element boundaries to satisfy compatibility of tangential slopes.

The properties of a selected deflection function may now be summarized as follows.

1. The function ideally contains a number of unknown parameters just equal to the total number of nodal degrees of freedom for the finite element.

2. The function must be capable of representing the rigid body motions of the finite element.

3. The function must be invariant with respect to any arbitrary geometric orientation of the finite element.

4. The matrix relating the nodal displacements of the finite element and the parameters of the selected deflection function must be non-singular.

5. The mathematical equations of the finite element formulation for any particular problem will possess more desirable convergence characteristics if the selected deflection function satisfies the conditions of compatibility of deflections and rotations of adjacent elements along common boundaries.

Once an appropriate displacement function has been selected, it is a relatively straightforward procedure to develop the governing nodal force-displacement relations for the element. From Eqs. (1.7), the strains, expressed in terms of the displacement of the middle surface, are

$$\left\{ \begin{array}{c} \epsilon_x \\ \epsilon_y \\ \gamma_{xy} \end{array} \right\} = \left\{ \begin{array}{c} -\mathcal{3}\,\dfrac{\partial^2 w}{\partial x^2} \\ -\mathcal{3}\,\dfrac{\partial^2 w}{\partial y^2} \\ -2\mathcal{3}\,\dfrac{\partial^2 w}{\partial x \partial y} \end{array} \right\} = \mathcal{3} \left\{ \begin{array}{c} \chi_x \\ \chi_y \\ \chi_{xy} \end{array} \right\} \qquad (9.6a)$$

which is expressed symbolically as

$$\{\epsilon\} = \mathcal{3}\{\chi\} \qquad (9.6b)$$

where $(\chi_x, \chi_y, \chi_{xy})$ are the plate curvatures and $\mathcal{3}$ is the normal distance measured from the middle surface. The curvatures may be expressed in matrix notation as

$$\{\chi\} = [Q]\{a\}. \qquad (9.7a)$$

The matrix $[Q]$ has the following form for the displacement function given by Eq. (9.1).

$$[Q] = \begin{bmatrix} 0 & 0 & 0 & -2 & 0 & 0 & -6x & -2y & 0 & 0 & -6xy & 0 \\ 0 & 0 & 0 & 0 & 0 & -2 & 0 & 0 & -2x & -6y & 0 & -6xy \\ 0 & 0 & 0 & 0 & -2 & 0 & 0 & -4x & -4y & 0 & -6x^2 & -6y^2 \end{bmatrix} \qquad (9.7b)$$

If we substitute Eq. (9.5a) into (9.7a), the symbolic expression for curvature becomes

$$\begin{aligned} \{\chi\} &= [Q]\{a\} \\ &= [Q][C^{-1}]\{u^n\} \\ &= [B]\{u^n\} \end{aligned} \qquad (9.8)$$

where $[B] = [Q][C^{-1}]$.

From Chapter 1, the stress resultants are

$$
\begin{Bmatrix} M_x \\ M_y \\ M_{xy} \end{Bmatrix} = \int_{-h/2}^{h/2} \mathfrak{z} \begin{Bmatrix} \sigma_x \\ \sigma_y \\ -\tau_{xy} \end{Bmatrix} d\mathfrak{z} \tag{9.9a}
$$

or, symbolically

$$
\{M\} = \int_{-h/2}^{h/2} \mathfrak{z}\{\sigma\} d\mathfrak{z} \tag{9.9b}
$$

where $(\sigma_x, \sigma_y, \tau_{xy})$ are the conventional normal and shearing stresses. The relationship between stress and strain from Eqs. (1.9) for homogeneous, isotropic, and linearly elastic materials has the following form.

$$
\begin{Bmatrix} \sigma_x \\ \sigma_y \\ \tau_{xy} \end{Bmatrix} = \frac{E}{1-\nu^2} \begin{bmatrix} 1 & \nu & 0 \\ \nu & 1 & 0 \\ 0 & 0 & \dfrac{1-\nu}{2} \end{bmatrix} \begin{Bmatrix} \epsilon_x \\ \epsilon_y \\ \gamma_{xy} \end{Bmatrix} \tag{9.10a}
$$

or, in matrix form

$$
\{\sigma\} = [\beta]\{\epsilon\}. \tag{9.10b}
$$

If we combine Eqs. (9.6a), (9.9a), and (9.10a) we obtain the following relationship between stress resultants or moments and curvatures.

$$
\{M\} = \left(\int_{-h/2}^{h/2} \mathfrak{z}^2 [\beta] \, d\mathfrak{z} \right) (\chi) \tag{9.11a}
$$

or

$$
\{M\} = [D]\{\chi\} \tag{9.11b}
$$

where

$$
[D] = \int_{-h/2}^{h/2} \mathfrak{z}^2 [\beta] \, d\mathfrak{z} \tag{9.11c}
$$

The matrix [D] for the isotropic case under consideration has the following simple form.

$$[D] = \frac{Eh^3}{12(1 - \nu^2)} \begin{bmatrix} 1 & \nu & 0 \\ \nu & 1 & 0 \\ 0 & 0 & \frac{1-\nu}{2} \end{bmatrix} \tag{9.12}$$

The moment-curvature relation given by Eq. (9.11b) may easily be extended to nonisotropic materials as well as layered media. This extension is treated in Section 9.5.

To continue the development, it is necessary to define the set of discrete nodal forces corresponding to the prescribed nodal degrees of freedom in Fig. 9.1.

$$\{F^n\} = \begin{Bmatrix} F_i \\ F_j \\ F_k \\ F_l \end{Bmatrix} \tag{9.13a}$$

where

$$\{F_m\} = \begin{Bmatrix} \overline{F}_w \\ M_x \\ M_y \end{Bmatrix}, \qquad m = i, j, k, l \tag{9.13b}$$

The set of element nodal forces given by Eq. (9.13a) must be statically equivalent to the stress resultants along the boundaries and the distributed load on the element. This static equivalence will be established by employing the principle of a stationary work function, according to Eq. (7.31). That is, the necessary and sufficient condition for the static equilibrium of a system is that the work function must have a stationary value. The work function for a plate element may be written in terms of (1) the internal strain energy of the element, (2) the work of the element nodal forces, and (3) the work of the external load p(x,y) on the element.

The internal strain energy, given by the integral of Eq. (7.11), is

$$U = \frac{1}{2} \int_V \{\epsilon\}^T \{\sigma\} \, dV. \tag{9.14a}$$

If we use Eqs. (9.6b) and (9.10b), Eq. (9.14a) becomes

$$U = \frac{1}{2} \int_V \mathfrak{z}^2 \{\chi\}^T \{\beta\}\{\chi\} \, dV. \tag{9.14b}$$

The strain energy may be reduced to the following integral over the area of the finite element by considering the case of a uniform thickness plate element, and using the moment-curvature relation given by Eq. (9.11a).

$$U = \frac{1}{2} \int_A \{\chi\}^T \{M\} \, dx \, dy \qquad (9.14c)$$

If we substitute Eq. (9.11b) and then (9.8) into Eq. (9.14c), we obtain

$$U = \frac{1}{2} \{u^n\}^T \left(\int_A [B]^T [D][B] \, dx \, dy \right) \{u^n\}. \qquad (9.15)$$

The work of the nodal forces is given by the product of the nodal forces and the displacements as follows.

$$W_1 = \{u^n\}^T \{F^n\} \qquad (9.16)$$

The work of the distributed load p(x,y) is

$$W_2 = \int_A w(x,y) \, p(x,y) \, dx \, dy. \qquad (9.17a)$$

If we substitute Eq. (9.2a) and then (9.5a) into Eq. (9.17a) we have

$$W_2 = \int_A [P][C^{-1}]\{u^n\} \, p(x,y) \, dx \, dy. \qquad (9.17b)$$

Since W_2 is a scalar quantity, we may take the transpose without changing its value.

$$W_2 = \int_A \{u^n\}^T [C^{-1}]^T [P]^T p(x,y) \, dx \, dy \qquad (9.17c)$$

The matrices $\{u^n\}$ and $[C^{-1}]^T$ contain constant elements that are independent of x and y; thus

$$W_2 = \{u^n\}^T [C^{-1}]^T \int_A [P]^T p(x,y) \, dx \, dy. \qquad (9.17d)$$

The final expression for the work function is

$$W = W_1 + W_2 - U$$

or

$$W = \{u^n\}^T\{F^n\} + \{u^n\}^T[C^{-1}] \int_A [P]^{\times}p(x,y)\,dx\,dy$$

$$- \frac{1}{2}\{u^n\}^T \left(\int_A [B]^T[D][B]\,dx\,dy \right)\{u^n\}. \tag{9.18}$$

For the work function to have a stationary value, we must have

$$\frac{\partial}{\partial u_m}(W) = 0 \qquad m = i, j, k, l. \tag{9.19}$$

If we perform the operation prescribed by Eq. (9.19) and rearrange terms, we obtain the following final expression that relates nodal displacements and corresponding nodal forces.

$$\{F^n\} = \left(\int_A [B]^T[D][B]\,dx\,dy \right)\{u^n\}$$

$$- [C^{-1}]^T \int_A [P]^T p(x,y)\,dx\,dy \tag{9.20a}$$

or

$$\{F^n\} = [K]\{u^n\} - \{f^n\} \tag{9.20b}$$

The finite element stiffness matrix [K] of dimension (12, 12) is

$$[K] = \int_A [B]^T[D][B]\,dx\,dy. \tag{9.21a}$$

Since $[B] = [Q][C]^{-1}$ from Eq. (9.8), the expression for the stiffness matrix may be written as

$$[K] = [C^{-1}]^T \left(\int_A [Q]^T[D][Q]\,dx\,dy \right)[C^{-1}]. \tag{9.21b}$$

The matrix $\{f^n\}$ of Eq. (9.20b) represents the external loads at the element nodes which are statically equivalent to the distributed load p(x,y) on the element.

Numerical values for the elements of the stiffness matrix [K] for the case of an orthotropic material are presented by Zienkiewicz and Cheung.[33] For the special case of an isotropic material, the stiffness matrix reduces to the expression:

$$[K] = \frac{Eh^3}{180\,ab(1-\nu^2)} [R] \left\{ [K_1] + [K_2] + \nu[K_3] + \frac{1-\nu}{2}[K_4] \right\} [R]$$

where

$$K_1 = \left(\frac{b}{a}\right)^2$$

60											
0	0					Symmetrical					
30	0	20									
30	0	15	60								
0	0	0	0	0							
15	0	10	30	0	20						
−60	0	−30	−30	0	−15	60					
0	0	0	0	0	0	0	0				
30	0	10	15	0	5	−30	0	20			
−30	0	−15	−60	0	−30	30	0	−15	60		
0	0	0	0	0	0	0	0	0	0	0	
15	0	5	30	0	10	−15	0	10	−30	0	20

$$K_2 = \left(\frac{a}{b}\right)^2$$

60											
−30	20					Symmetrical					
0	0	0									
−60	30	0	60								
−30	10	0	30	20							
0	0	0	0	0	0						
30	−15	0	−30	−15	0	60					
−15	10	0	15	5	0	−30	20				
0	0	0	0	0	0	0	0	0			
−30	15	0	30	15	0	−60	30	0	60		
−15	5	0	15	10	0	−30	10	0	30	20	
0	0	0	0	0	0	0	0	0	0	0	0

$$K_3 =$$

30											
−15	0					Symmetrical					
15	−15	0									
−30	0	−15	30								
0	0	0	15	0							
−15	0	0	15	15	0						
−30	15	0	30	0	0	30					
15	0	0	0	0	0	−15	0				
0	0	0	0	0	0	−15	15	0			
30	0	0	−30	−15	0	−30	0	15	30		
0	0	0	−15	0	0	0	0	0	15	0	
0	0	0	0	0	0	15	0	0	−15	−15	0

$$K_4 = \begin{bmatrix}
84 \\
-6 & 8 \\
6 & 0 & 8 & & & \text{Symmetrical} \\
-84 & 6 & -6 & 84 \\
-6 & -2 & 0 & 6 & 8 \\
-6 & 0 & -8 & 6 & 0 & 8 \\
-84 & 6 & -6 & 84 & 6 & 6 & 84 \\
6 & -8 & 0 & -6 & 2 & 0 & -6 & 8 \\
6 & 0 & -2 & -6 & 0 & 2 & -6 & 0 & 8 \\
84 & -6 & 6 & -84 & -6 & -6 & -84 & 6 & 6 & 84 \\
6 & 2 & 0 & -6 & -8 & 0 & -6 & -2 & 0 & 6 & 8 \\
-6 & 0 & 2 & 6 & 0 & -2 & 6 & 0 & -8 & -6 & 0 & 8
\end{bmatrix}$$

$$[R] = \begin{bmatrix}
[r] & [0] & [0] & [0] \\
[0] & [r] & [0] & [0] \\
[0] & [0] & [r] & [0] \\
[0] & [0] & [0] & [r]
\end{bmatrix}$$

$$[r] = \begin{bmatrix}
1 & 0 & 0 \\
0 & b & 0 \\
0 & 0 & a
\end{bmatrix}.$$

This element stiffness matrix has been thoroughly tested and in general provides satisfactory accuracy and convergence.

9.3 TRIANGULAR PLATE ELEMENTS

In Fig. 9.3 a flat plate is shown subdivided into a set of triangular finite elements. The geometrical position of a triangular element is determined by the three nodal points (i,j,k) and the straight line boundaries. We again assume that each nodal point of an element has a transverse deflection w and two components of rotation, the three of which are referred to as a set of nodal displacements. We also assume that a transverse force and two components of moment act at each nodal point, and that these three loads are referred to as a set of nodal forces. The nodal displacements and nodal forces are assumed to be positive in the directions shown in Fig. 9.2. The coordinate set (X,Y,Z) again is called the global set, while the set (x,y,z) is referred to as the local set for the finite element under consideration.

In this section we begin with a selected deflection function and then

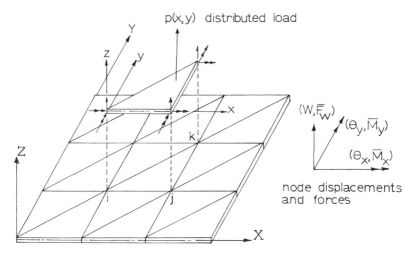

Fig. 9.2.   Triangular plate elements

develop the relationship between the three sets of nodal displacements and the three sets of nodal forces for a triangular plate element. Again, this relationship is presented in the form of a matrix which we refer to as a stiffness matrix.

The task of selecting a suitable displacement function for the triangular element is somewhat more difficult than for the rectangular element. The simplest polynomial expression for the deflection of the middle surface which satisfies the majority of the criteria established in Section 9.2 is

$$w(x,y) = a_1 + a_2x + a_3y + a_4x^2 + a_5xy + a_6y^2$$
$$+ a_7x^3 + a_8x^2y + a_9xy^2 + a_{10}y^3. \qquad (9.22)$$

For this cubic function to be invariant with respect to any arbitrary orientation of the finite element in the x-y plane, it must be a complete cubic function; thus, it must contain all ten terms. Paradoxically, there are but nine degrees of nodal freedom for a triangular element. Investigators have been prompted to attempt several different alterations of Eq. (9.22). Among these alterations are (1) letting $a_8 = 0$, (2) letting $a_9 = 0$, and (3) letting $a_8 = a_9$. All three of these arbitrary choices violate the required invariance of the displacement function, and for some geometric configurations the required matrix inversion analogous to Eq. (9.5a) will prove to be singular. Several investigators have compared solutions using these three forms to closed form solutions of selected classical problems. Convergence to the exact solution with decreasing element size is never assured, and the finite element solutions are at best marginal.

This difficulty has prompted a considerable effort to find a suitable displacement function for the triangular element. Gallagher[35] has presented an excellent review of the research efforts concentrated on defining additional degrees of freedom for the element. These degrees of freedom include the already mentioned deflection and rotations at each node, supplemented by the three curvatures at each nodal point and the rotations at the midpoints of each boundary. Such formulations are defined as mixed or hybrid, and contain anywhere from 9 to 21 degrees of freedom. A complete fifth-order polynomial is normally used to model the element. These additional terms do improve the convergence characteristics; however, they also impose a considerable increase and complexity in the labor of solving a problem.

Another approach to the triangular element problem has been the subelement concept advanced by Clough and Tocher.[36] In this approach the triangular element is subdivided into three subelements, A, B, and C, as shown in Fig. 9.3.

Point o is any point convenient to the interior of the element, usually the centroid. Points (p,q,r) are midpoints on the three interior boundaries, and $(\theta_p, \theta_q, \theta_r)$ represent the rotations or slopes tangent to the interior boundaries at these points. Each subelement is identified with

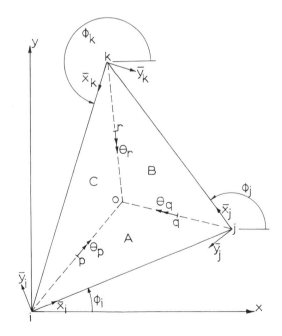

Fig. 9.3. Triangular subelements

a local $(\bar{x},\bar{y},\bar{z})$ system of coordinates with the $\bar{x}$ axis taken along the exterior boundary. The selected displacement functions for each of the three subelements are complete third-degree polynomials with the x^2y term missing in each case:

$$\bar{w}_A = a_1 + a_2\bar{x}_i + a_3\bar{y}_i + \ldots + a_7\bar{x}_i^3 + a_8\bar{x}_i\bar{y}_i^2 + a_9\bar{y}_i^3 \quad (9.23a)$$

$$\bar{w}_B = b_1 + b_2\bar{x}_i + b_3\bar{y}_i + \ldots + b_7\bar{x}_i^3 + b_8\bar{x}_i\bar{y}_i^2 + b_9\bar{y}_i^3 \quad (9.23b)$$

$$\bar{w}_C = c_1 + c_2\bar{x}_i + c_3\bar{y}_i + \ldots + c_7\bar{x}_i^3 + c_8\bar{x}_i\bar{y}_i^2 + c_9\bar{y}_i^3. \quad (9.23c)$$

Each polynomial is capable of representing rigid body motion of the subelement by virtue of the first three terms, as discussed in Section 9.2.

We see that these polynomials are expressed in terms of twenty-seven independent parameters, a_i, b_i, c_i, where $i = 1, 2, \ldots, 9$. In the process of formulating the stiffness matrix for a triangular element we must determine the twenty-seven parameters in terms of the nodal displacements. To determine these parameters, we require the deflections of adjacent subelements to be compatible (equal) at the common nodal points. We make similar requirements for rotations. We also require the tangential slope of a subelement to be equal to the tangential slope of the adjacent subelement at the midpoint of the common boundary. As an example, θ_p of subelement A is set equal to θ_p of subelement C at Point p. Desirable properties of the selected polynomials involve (1) compatibility of deflections and rotations along common boundaries of adjacent elements, and (2) compatibility of deflections and rotations along common boundaries of adjacent subelements.

(1) **Adjacent Elements.** Let us consider a common boundary line of two adjacent elements, such as line i-j in Fig. 9.3. This boundary is common to the element shown in the figure and is also common to the adjacent element, which is not shown. The general expression for the deflection of each of the two adjacent elements along this boundary is a cubic. Thus, as discussed in Section 9.2, the deflections and normal rotations of the two adjacent elements along this common boundary are compatible. The general expression for the tangential rotation θ_x of each of the two adjacent elements along this common boundary is a linear function, due to the omission of the $\bar{x}^2\bar{y}$ term in each subelement displacement function. The value of the tangential rotation at each of the two nodal points i and j provides sufficient information to determine the two constants that define the resulting linear function. Since this information is identical for the adjacent elements, we conclude that tangential rotations of the adjacent elements along the common bound-

ary are compatible for the triangular element. This tangential rotation compatibility is in contrast with the lack of tangential rotation compatibility for the rectangular element discussed in Section 9.2.

(2) Adjacent Subelements. Consider the common boundary i-o, Fig. 9.3, between the adjacent subelements A and C. The general expression for the deflection of each subelement along this common boundary is a cubic. Thus, again as discussed in Section 9.2, the deflections and normal rotations of the two adjacent subelements along this common boundary are compatible. The general expression for tangential rotation of each subelement along this common boundary is a quadratic, which is defined by three constants. The value of tangential rotation at each of the three points (i,p,o) provides the necessary information to determine these constants. Since these values of rotation are identical for the two adjacent subelements, we conclude that we have, at the least, compatibility of tangential rotations at the midpoints and extremities of the common boundaries of adjacent subelements.

A set of nodal displacements may be expressed in either the subelement coordinate system $(\bar{x},\bar{y},z)$ or the element coordinate system (x,y,z). The displacements of a nodal point in terms of the $(\bar{x},\bar{y},z)$ coordinates are related to the displacements of that point in terms of the (x,y,z) coordinates by the equation:

$$\begin{Bmatrix} w \\ \theta_x \\ \theta_y \end{Bmatrix} = \begin{bmatrix} 1 & 0 & 0 \\ 0 & \cos\phi & -\sin\phi \\ 0 & \sin\phi & \cos\phi \end{bmatrix} \begin{Bmatrix} \bar{w} \\ \theta_{\bar{x}} \\ \theta_{\bar{y}} \end{Bmatrix} \tag{9.24a}$$

or

$$\{u_m\} = [\Phi_m]\{\bar{u}_m\} \qquad m = \text{nodal point.} \tag{9.24b}$$

We begin the formulation of the stiffness matrix for the triangular element by determining the expression for the twenty-seven independent parameters of Eqs. (9.23). If we refer to Eqs. (9.3) and (9.4) and make use of Eq. (9.24a), we may express the nodal displacements of subelement A in the (x,y,z) coordinate system as follows.

$$\{u_i\} = [A_i]\{a\} \tag{9.25a}$$

$$\{u_j\} = [A_j]\{a\} \tag{9.25b}$$

$$\{u_o\} = [A_o]\{a\} \tag{9.25c}$$

$$\theta_p = [A_p]\{a\} \tag{9.25d}$$

$$\theta_q = [A_q]\{a\} \tag{9.25e}$$

The matrices $[A_i]$, $[A_j]$, and $[A_o]$, which are associated with the nodal points (i,j,o) respectively, are each of dimension (3,9). The matrices $[A_p]$ and $[A_q]$, which are associated with the nodal points (p,q) respectively, are each of dimension (1,9). Remember, the nodal displacements $(u_i, u_j, u_o, \theta_p, \theta_q)$ are expressed in the (x,y,z) coordinate system as a result of applying Eq. (9.24a). The nodal displacements of subelements B and C in the (x,y,z) coordinate system also may be expressed in matrix form as follows.

$$\{u_j\} = [B_j]\{b\} \tag{9.26a}$$

$$\{u_k\} = [B_k]\{b\} \tag{9.26b}$$

$$\{u_o\} = [B_o]\{b\} \tag{9.26c}$$

$$\theta_q = [B_q]\{b\} \tag{9.26d}$$

$$\theta_p = [B_r]\{b\} \tag{9.26e}$$

$$\{u_k\} = [C_k]\{c\} \tag{9.27a}$$

$$\{u_i\} = [C_i]\{c\} \tag{9.27b}$$

$$\{u_o\} = [C_o]\{c\} \tag{9.27c}$$

$$\theta_r = [C_r]\{c\} \tag{9.27d}$$

$$\theta_p = [C_p]\{c\} \tag{9.27e}$$

If we require that the nodal displacements of the common nodes of adjacent subelements be equal, and that the tangential rotations of the midpoints of the common boundaries of adjacent subelements be equal, we obtain the following equations.

$$[A_i]\{a\} = [C_i]\{c\} \tag{9.28a}$$

$$[A_j]\{a\} = [B_j]\{b\} \tag{9.28b}$$

$$[B_k]\{b\} = [C_k]\{c\} \tag{9.28c}$$

$$[A_o]\{a\} = [B_o]\{b\} \qquad (9.28d)$$

$$[B_o]\{a\} = [C_o]\{c\} \qquad (9.28e)$$

$$[A_p]\{a\} = [C_p]\{c\} \qquad (9.28f)$$

$$[A_q]\{a\} = [B_q]\{b\} \qquad (9.28g)$$

$$[B_r]\{b\} = [C_r]\{c\} \qquad (9.28h)$$

Equations (9.28), along with Eqs. (9.25a), (9.26a), and (9.27a) represent 27 independent equations in the 27 parameters (a,b,c) of the displacement functions. These 27 equations may be expressed in the following matrix form.

$$
\underset{(27,1)}{\begin{Bmatrix} u_i \\ u_j \\ u_k \\ 0 \\ 0 \\ 0 \\ 0 \\ 0 \\ 0 \\ 0 \\ 0 \end{Bmatrix}}
=
\underset{(27,27)}{\begin{bmatrix}
A_i & 0 & 0 \\
0 & B_j & 0 \\
0 & 0 & C_k \\
A_i & 0 & -C_i \\
A_j & -B_j & 0 \\
0 & B_k & -C_k \\
A_o & -B_o & 0 \\
0 & B_o & -C_o \\
A_p & 0 & -C_p \\
A_q & -B_q & 0 \\
0 & B_r & -C_r
\end{bmatrix}}
\underset{(27,1)}{\begin{Bmatrix} a \\ b \\ c \end{Bmatrix}} \qquad (9.29)
$$

The (27,27) array of Eq. (9.29) may then be inverted to determine the 27 parameters (a,b,c). The task of verifying the non-singularity of this matrix is formidable; however, no investigator has reported any computational failure for a wide class of problems. The resulting inverse operation may be represented symbolically as

$$
\begin{Bmatrix} a \\ b \\ c \end{Bmatrix}
=
\begin{bmatrix} R_1 \\ R_2 \\ R_3 \end{bmatrix} \{u^n\} . \qquad (9.30)
$$

The remainder of the procedure for the formulation of the stiffness matrix for the triangular element follows the same procedure presented in Section 9.2 for the rectangular element. The resulting stiffness matrix of dimension (9,9) for the triangular element is

$$[K] = [R_1]^T \left(\int_{A_1} [Q_A]^T [D][Q_A] \, d\bar{x}_i \, d\bar{y}_i \right) [R_1]$$
$$+ [R_2]^T \left(\int_{A_2} [Q_B]^T [D][Q_B] \, d\bar{x}_j \, d\bar{y}_j \right) [R_2]$$
$$+ [R_3]^T \left(\int_{A_3} [Q_C]^T [D][Q_C] \, d\bar{x}_k \, d\bar{y}_k \right) [R_3] \qquad (9.31)$$

where A_1, A_2, and A_3 are the areas of subelements A, B, and C respectively. The matrix [D] is the matrix defined by Eq. (9.12). The matrices $[Q_A]$, $[Q_B]$, and $[Q_C]$, which correspond to the subelements A, B, and C respectively, are analogous to the matrix given by Eq. (9.7b). The matrix $[Q_A]$, which is for subelement A, has the following form.

$$[Q_A] = \begin{bmatrix} 0 & 0 & 0 & -2 & 0 & 0 & -6\bar{x}_i & 0 & 0 \\ 0 & 0 & 0 & 0 & 0 & -2 & 0 & -2\bar{x}_i & -6\bar{y}_i \\ 0 & 0 & 0 & 0 & -2 & 0 & 0 & -4\bar{y}_i & 0 \end{bmatrix} \qquad (9.32a)$$

where

$$\left\{ \begin{array}{c} -\dfrac{\partial^2 \bar{w}_A}{\partial \bar{x}_i^2} \\[2mm] -\dfrac{\partial^2 \bar{w}_A}{\partial \bar{y}_i^2} \\[2mm] -2\dfrac{\partial^2 \bar{w}_A}{\partial \bar{x}_i \, \partial \bar{y}_i} \end{array} \right\} = [Q_A]\{a\} \qquad (9.32b)$$

The corresponding expressions for $[Q_B]$ and $[Q_C$ are similar to Eq. (9.32a) and may be determined easily. It may be shown numerically that the stiffness matrix [K] of Eq. (9.31) is invariant with respect to any arbitrary orientation of the finite element.

Clough and Felippa[37] have published a refinement of the subelement triangular element stiffness matrix. The refinement utilizes the full third-order polynomial comprised of all ten terms given by Eq. (9.22), and is accomplished by adding an additional three degrees of freedom to the element. These additional three degrees of freedom are the slopes at the midpoints of the exterior boundaries. The resulting 12-degree-of-freedom element has slightly better convergence characteristics than the nine-degree-of-freedom triangular element presented in this section.

9.4 COMPARISON OF RESULTS

In this section deflections obtained by the following three techniques are compared:

1. a finite element analysis using the rectangular element discussed in Section 9.2;

2. a finite element analysis using the triangular element discussed in Section 9.3;

3. a closed form series solution.

For the comparison, we consider the simply supported plate under a single centrally applied concentrated load in Fig. 9.4.[35]

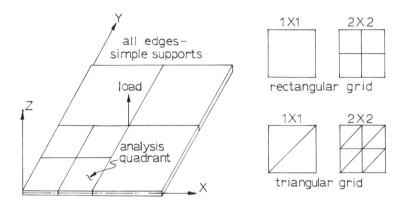

Fig. 9.4.   Simply supported plate with a centrally applied concentrated load

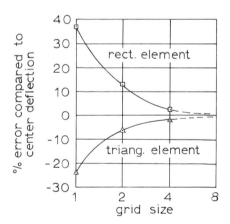

Fig. 9.5.   Comparison results

Figure 9.5 illustrates the convergence of the value of the center deflection computed by the finite element techniques to the value of the center deflection computed by the series solution. Note that, although the number of elements for a 2 × 2 triangular grid is considerably larger than the number of elements for a 2×2 rectangular grid, the number of degrees of freedom is equal.

9.5 ORTHOTROPIC PLATES

The finite element formulations discussed in the previous sections are extended to orthotropic plates by using the appropriate orthotropic stress-strain relations. For an orthotropic plate element in which the local coordinate directions coincide with the principal directions of orthotropy, we rewrite Eqs. 6.2 in matrix form as follows.

$$\{\sigma\} = [\beta_o]\{\epsilon\} \tag{9.33a}$$

where

$$[\beta_o] = \begin{bmatrix} \dfrac{E'_x}{1 - \nu_x \nu_y} & \dfrac{\nu_y E'_x}{1 - \nu_x \nu_y} & 0 \\ \dfrac{\nu_x E'_y}{1 - \nu_x \nu_y} & \dfrac{E'_y}{1 - \nu_x \nu_y} & 0 \\ 0 & 0 & G \end{bmatrix} \tag{9.33b}$$

In many instances, the principal directions of orthotropy are not oriented along the local coordinate axes. Instead, they are oriented along an arbitrary set of axes as shown in Fig. 9.6.

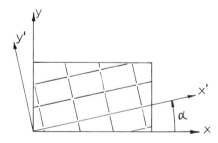

Fig. 9.6. Transversely orthotropic material

The appropriate orthotropic stress-strain relations corresponding to (the (x',y') axes are given by

$$\{\sigma'\} = [\beta'_o]\{\epsilon'\} \tag{9.35}$$

where

$$\{\sigma'\} = \left\{ \begin{array}{c} \sigma_{x'} \\ \sigma_{y'} \\ \tau_{x'y'} \end{array} \right\}$$

$$\{\epsilon'\} = \left\{ \begin{array}{c} \epsilon_{x'} \\ \epsilon_{y'} \\ \gamma_{x'y'} \end{array} \right\}$$

$$[\beta_o] = \begin{bmatrix} \dfrac{E'_{x'}}{1 - \nu_{x'}\nu_{y'}} & \dfrac{\nu_{y'} E'_{x'}}{1 - \nu_{x'}\nu_{y'}} & 0 \\ \dfrac{\nu_{x'} E'_{y'}}{1 - \nu_{x'}\nu_{y'}} & \dfrac{E'_{y'}}{1 - \nu_{x'}\nu_{y'}} & 0 \\ 0 & 0 & G \end{bmatrix}.$$

The strains corresponding to the (x',y',z) axes are related to the strains corresponding to the (x,y,z) axes by the following strain transformation matrix.[5]

$$\{\epsilon'\} = [A]\{\epsilon\} \tag{9.36a}$$

where

$$[A] = \begin{bmatrix} \cos^2 \alpha & \sin^2 \alpha & \sin \alpha \cos \alpha \\ \sin^2 \alpha & \cos^2 \alpha & -\sin \alpha \cos \alpha \\ -2 \sin \alpha \cos \alpha & 2 \sin \alpha \cos \alpha & \cos^2 \alpha - \sin^2 \alpha \end{bmatrix} \tag{9.36b}$$

Similarly, the stresses in the two coordinate systems are related by[5]

$$\{\sigma\} = [A]^T\{\sigma'\}. \tag{9.37}$$

Thus, from Eqs. (9.35), (9.36a), and (9.37) we obtain

$$\{\sigma\} = [A]^T[\beta'_o][A]\{\epsilon\} = [\beta_o]\{\epsilon\} \tag{9.38a}$$

where

$$[\beta_o] = [A]^T[\beta'_o][A]. \tag{9.38b}$$

We recall from the theory of orthotropic plates that the orthotropic constants are assumed to be invariant in the direction perpendicular to the plane of the plate. Consequently, we may integrate Eq. (9.11c) and obtain

$$[D] = \int_{-h/2}^{h/2} \mathcal{Z}^2[\beta_o] \, d\mathcal{Z} = \frac{h^3}{12} [\beta_o].$$ (9.39)

The stiffness matrices for an orthotropic rectangular element and an orthotropic triangular element are obtained by substituting the expression for [D] given by Eq. (9.39) into Eqs. (9.21b) and (9.31) respectively.

It is also possible to develop the moment-curvature relationship, [D], for layered plate elements. An example of this type of plate element is seen in Fig. 9.7.

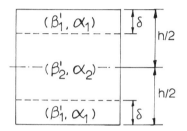

Fig. 9.7. Composite plate structure

In this example $[\beta_{01}']$ and $[\beta_{02}']$ define the stress-strain relations of each layer. It should be evident that each individual layer can be composed of either an orthotropic or an isotropic material. We may generalize our example by assuming that each of the layers contains an orthotropic material. (This includes isotropic materials as a special case of orthotropy.) We generalize further by assuming that the directions of principal orthotropy for Layers 1 and 2 form angles with the local coordinate axes of α_1 and α_2 respectively. Then we write

$$[\beta_{01}] = [A_1]^T[\beta_{01}'][A_1]$$ (9.40a)

and

$$[\beta_{02}] = [A_2]^T[\beta_{02}'][A_2]$$ (9.40b)

where the transformation matrices $[A_1]$ and $[A_2]$ are evaluated for the angles α_1 and α_2 respectively.

Finally, by assuming complete continuity of shear strain between the layers, we can integrate Eq. (9.11) and establish the following moment-curvature relationship for our example.

$$[D] = 2 \int_0^{h/2-\delta} \tilde{z}^2[\beta_2] \, d\tilde{z} + 2 \int_{h/2-\delta}^{h/2} \tilde{z}^2[\beta_1] \, d\tilde{z} \qquad (9.41)$$

Again, the stiffness matrices for layered rectangular and triangular elements are obtained by substituting the expression [D] given by Eq. (9.41) into Eqs. (9.21b) and (9.31) respectively. If the individual layers are orthotropic, then errors are introduced into the solution if the layers are not symmetric about the composite middle surface. This occurs because the neutral surfaces corresponding to bending about the directions of principal orthotropy will not coincide.

9.6 LARGE DEFLECTIONS

In this section the analysis of large lateral deflections of thin plates is divided into two categories.

1. Terms of higher order are included in the strain-displacement relations to account for the existence of middle surface strains. Middle surface strains were negligible in small displacement theory according to Assumption 1 in Section 1.1.

2. Lateral deflections are determined for plates of materials with nonlinear stress-strain relations. The nonlinearity of the stress-strain relations is often encountered in the plastic range as a result of large deflections.

Category 1. In this category we discuss plate deflections that are no longer small in comparison to the plate thickness, but are still considered small when compared to the other plate dimensions. Our previous discussion must now be extended to include the effect of strains in the middle surface of the plate for these large deflection problems. These middle surface strains will develop whenever the plate is deformed into a nondevelopable surface, and they become sizable when the deflections are large. Because of the inclusion of middle surface strains, we may now include in-plane applied forces as well as lateral loads.

The general strain-displacement relations for large deflections are[1]

$$\epsilon_x = \frac{\partial u}{\partial x} + \frac{1}{2} \left(\frac{\partial w}{\partial x} \right)^2 \qquad (9.42a)$$

$$\epsilon_y = \frac{\partial v}{\partial y} + \frac{1}{2} \left(\frac{\partial w}{\partial y} \right)^2 \qquad (9.42b)$$

$$\gamma_{xy} = \frac{\partial u}{\partial y} + \frac{\partial v}{\partial x} + \frac{\partial w}{\partial x}\frac{\partial w}{\partial y}. \qquad (9.42c)$$

For the case of small deflection theory, we assumed that the square of the small displacement gradient terms could be neglected; however, we now wish to include these nonlinear strain terms to more nearly satisfy the case of large deflections. As a starting point, we assume an initial system of in-plane stresses $\{\sigma^0\}$ in the plate.

$$\{\sigma^0\} = \left\{ \begin{array}{c} \sigma_x^0 \\ \sigma_y^0 \\ \tau_{xy}^0 \end{array} \right\} \qquad (9.43)$$

We also assume that these in-plane stresses will remain unchanged during any bending of the plate. This assumption is widely used in the classical treatment of this problem.[1]

With these assumptions, we may express the strain relations given by Eqs. (9.42) as follows.

$$\epsilon_x = -\zeta\frac{\partial^2 w}{\partial x^2} + \epsilon_x^0 \qquad (9.44a)$$

$$\epsilon_y = -\zeta\frac{\partial^2 w}{\partial y^2} + \epsilon_y^0 \qquad (9.44b)$$

$$\gamma_{xy} = -2\zeta\frac{\partial^2 w}{\partial x\,\partial y} + \gamma_{xy}^0 \qquad (9.44c)$$

where

$$\epsilon_x^0 = \frac{1}{2}\left(\frac{\partial w}{\partial x}\right)^2 \qquad (9.44d)$$

$$\epsilon_y^0 = \frac{1}{2}\left(\frac{\partial w}{\partial y}\right)^2 \qquad (9.44e)$$

$$\gamma_{xy}^0 = \frac{\partial w}{\partial x}\frac{\partial w}{\partial y} \qquad (9.44f)$$

From Eqs. (9.6a) we interpret the first terms of Eqs. (9.44a), (9.44b), and (9.44c) as the strain from bending; we interpret the second terms as the strain terms related to the set of initial stresses $\{\sigma^0\}$. These second terms are assumed to be constant over the thickness of the plate. The strain energy per unit volume corresponding to these constant middle surface stresses and strains may be written as

$$\sigma_x^0 \epsilon_x^0 + \sigma_y^0 \epsilon_y^0 + \tau_{xy}^0 \gamma_{xy}^0. \tag{9.45}$$

Since the bending strains and the middle surface strains are assumed to be independent, we modify the total strain energy of the finite element given by Eq. (9.14c) to the following form.

$$U = \frac{1}{2} \int_{Area} \{\chi\}^T \{M\} \, dx \, dy$$

$$+ \frac{h}{2} \int_{Area} \begin{Bmatrix} \dfrac{\partial w}{\partial x} \\ \dfrac{\partial w}{\partial y} \end{Bmatrix}^T \begin{bmatrix} \sigma_x^0 & \tau_{xy}^0 \\ \tau_{xy}^0 & \sigma_y^0 \end{bmatrix} \begin{Bmatrix} \dfrac{\partial w}{\partial x} \\ \dfrac{\partial w}{\partial y} \end{Bmatrix} \, dx \, dy \tag{9.46}$$

The slopes $\partial w / \partial x$ and $\partial w / \partial y$ are related to the nodal displacements of the finite element through the assumed displacement function

$$\begin{Bmatrix} \dfrac{\partial w}{\partial x} \\ \dfrac{\partial w}{\partial y} \end{Bmatrix} = [S]\{a\}. \tag{9.47a}$$

If we substitute Eq. (9.5a) into (9.47a) we obtain

$$\begin{Bmatrix} \dfrac{\partial w}{\partial x} \\ \dfrac{\partial w}{\partial y} \end{Bmatrix} = [S][C^{-1}]\{u^n\} = [H]\{u^n\} \tag{9.47b}$$

where

$$[H] = [S][C^{-1}].$$

We substitute Eq. (9.47b) into (9.46) and obtain the following form for the strain energy in the finite element.

$$U = \frac{1}{2} \int_{Area} \{\chi\}^T \{M\} \, dx \, dy$$

$$+ \frac{h}{2} \{u^n\}^T \left(\int_{Area} [H]^T [\sigma^0][H] \, dx \, dy \right) \{u_n\} \tag{9.48a}$$

where

$$[\sigma^0] = \begin{bmatrix} \sigma_x^0 & \tau_{xy}^0 \\ \tau_{xy}^0 & \sigma_y^0 \end{bmatrix} \tag{9.48b}$$

If we apply the principle of a stationary work function according to Section 7.5, the following stiffness relation similar to Eq. (9.20b) is obtained.

$$\{F^n\} = [K]\{u^n\} + [K_G]\{u^n\} = [[K] + [K_G]]\{u^n\} \tag{9.49}$$

where $[K]$ is given by Eq. (9.21b), and $[K_G]$, the geometrical stiffness matrix, is given by

$$[K_G] = h[C^{-1}]^T \left(\int_{\text{Area}} [S]^T [\sigma^0][S] \, dx \, dy \right) [C^{-1}]. \tag{9.50}$$

This matrix depends on the numerical values of the constant middle surface stresses (σ_x^0, σ_y^0, τ_{xy}^0), and may be written in the simplified form

$$[K_G] = \sigma_x^0 [K_{G_x}] + \sigma_y^0 [K_{G_y}] + \gamma_{xy}^0 [K_{G_{xy}}] \tag{9.51a}$$

where $[K_{G_x}]$, for example, would be evaluated as

$$[K_{G_x}] = h[C^{-1}]^T \left(\int_{\text{Area}} [S]^T \begin{bmatrix} 1 & 0 \\ 0 & 0 \end{bmatrix} [S] \, dx \, dy \right) [C^{-1}] \tag{9.51b}$$

with similar evaluations for $[K_{G_y}]$ and $[K_{G_{xy}}]$.

We observe from Eq. (9.49) that the stiffness of a finite element is modified as a result of the middle surface strains by the addition of $[K_G]$. In the case of tensile middle surface stresses, the geometrical stiffness is additive; however, when the middle surface stresses are compressive, the total stiffness decreases.

The task of solving a problem generally proceeds as follows. In the absence of any reasonable estimates for the middle surface stresses, we assume their initial state as zero, which leads to a conventional finite element problem as described in the preceding sections. From the resulting set of nodal displacements determined from this conventional analysis, we calculate the set of slopes given by Eq. (9.47b) at some representative point in the finite element. For example, we might select the value of the slopes at the area centroid of the element to represent the slopes of the element. We then calculate the initial middle surface strains with Eqs. (9.44d), (9.44e), and (9.44f) and the corresponding middle surface stresses with the appropriate stress-strain relationships.

With this information, the geometrical stiffness matrix, Eq. (9.51a), is numerically evaluated and added to the bending stiffness matrix as given in Eq. (9.49). We again solve the finite element problem using the modified element stiffness properties as given in Eq. (9.49). Proceeding in this way, we may iteratively resolve the problem to attain satisfactory convergence of the middle surface stresses, $(\sigma_x^0, \sigma_y^0, \tau_{xy}^0)$. Two or three iterations are generally sufficient for most practical problems.

This method is particularly well suited for the case of a laterally loaded plate which is simultaneously loaded with known in-plane forces along the edges. The stresses resulting from these known in-plane forces are used as the set of initial in-plane stresses $\{\sigma^0\}$ for the first iteration. This method also has been used in the study of plate buckling.[33]

Category 2. Large lateral deflections of a plate often result in large strains. Consequently, the stresses in the plate might not be linearly related to the corresponding strains. We can determine the lateral deflections, stresses, and strains for such a plate by continually modifying the material properties in an iterative process. The first step in this process is to determine the strains in each element by performing a conventional linearly elastic finite element analysis in which the values of the material properties correspond to the linear portions of the appropriate stress-strain curves. Comparison of these element strains with the appropriate stress-strain curves may reveal that the strains in some regions of the plate are not within the linear portion of the curve as shown in Fig. 9.8. Thus, these strains, as determined by the linear analysis, are not exact. If we are to use a linear analysis we must adjust the material properties of each element that contains strains which are not within the linear portion of the stress-strain curve. There are many techniques with which we can determine the adjusted material properties for these elements. The state of the art for determining these adjusted material properties is in an early stage of development, and no particular technique has received universal acceptance.

However, to introduce the reader to the type of techniques which can be used, let's assume that we have numbered the elements and the Nth element contains strains that are not within the linear portion of the stress-strain curve. We can obtain the set of outer surface strains (strains on the outer surface are maximum since strain varies linearly with $\mathfrak{z}$) corresponding to the point at the centroid of the area of this element from our initial linear finite element analysis. From this set of strains we determine the maximum absolute value, ϵ_{N1}, of the principal strains at the centroid of the element. The subscript N refers to the Nth element and the subscript 1 refers to the first approximation. Then we

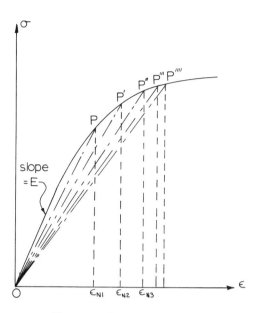

Fig. 9.8. Stress-strain curve

locate ϵ_{N1} on the appropriate stress-strain curve as illustrated in Fig. 9.8 and determine the slope of the line OP. We identify this slope as E_{N1} and replace the original value of E for element N by the value E_{N1}. In this example we assume that ν remains constant. If the variation of strain across an individual element is large enough to create question about the validity of using the strain at the centroid as an indicator of the total element strain, the problem could be reformulated with smaller elements. The same procedure is repeated to determine adjusted material properties for all elements containing strains outside of the linear portion of the stress-strain curves.

Next we determine revised values of the element strains with a second linear finite element analysis in which we use the adjusted material properties of the elements. Again we determine the maximum absolute value of the principal strain at the outer surface of the element at the centroid. We adjust the material properties of all elements for which these strains are outside the linear portion of the stress–strain curves. As an example, we replace the material property E_{N1} by E_{N2} where E_{N2} is the slope of line OP' in Fig. 9.8. The preceding process is repeated again and again, as illustrated in Fig. 9.8, until the investigator is satisfied that the maximum centroidal strain of each element converges.

In the analysis of plates for which the elastic region of the stress-strain curve is nonlinear, the preceding iterative procedure can be used with the following two changes.

1. An initial estimate of E must be made since the stress-strain curve has no linear regions.

2. All element strains will correspond to nonlinear regions on the stress-strain curve; therefore, during the iteration process the material properties must be adjusted for all the elements.

The analyses of plates which involve nonlinear stress-strain behavior also can be conducted with an incremental technique in which the stress-strain curves are approximated by a series of straight lines.[33]

Problems

39. Develop the elements of the matrix C given by Eq. (9.4a) for a rectangular element of aspect ratio a/b, and demonstrate that Eq. (9.5b) is the inverse of this array.

40. Consider a rectangular finite element with displacement degrees of freedom as shown in Fig. 9.9.

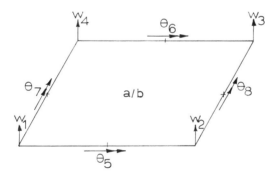

Fig. 9.9

a. Using a displacement function of the form

$$w = a_1 + a_2x + a_3y + a_4x^2 + a_5xy + a_6y^2 + a_7x^3 + a_8y^3$$

develop the governing force-displacement equations for the element, assuming an isotropic material. The rotational degrees of freedom are to be taken at the midpoints of the element boundaries.

b. Discuss how these elements would be combined to form the stiffness matrix for a plate structure.

c. Briefly summarize an analogous model for the case of a triangular finite element.

41. Develop the displacement compatibility matrix given by Eqs. (9.28) for the case of a triangular finite element containing a 90° angle.

REFERENCES

1. Timoshenko, S., and Woinowsky-Krieger, S. *Theory of Plates and Shells.* New York: McGraw-Hill Book Co., 1959.
2. Timoshenko, S., and Goodier, J. N. *Theory of Elasticity.* New York: McGraw-Hill Book Co., 1951.
3. Love, A. E. H. *A Treatise on the Mathematical Theory of Elasticity.* New York: Dover Publications, 1944.
4. Reissner, E. "The Effect of Transverse Shear Deformation on the Bending of Elastic Plates." *Journal of Applied Mechanics* 12 (1945): A-68.
5. Wang, Chi-Teh. *Applied Elasticity.* New York: McGraw-Hill Book Co., 1953.
6. Kirchhoff, G. "Über das Gleichgewicht und die Bewegung einer elastishen Scheibe." *Journal fuer die reine und angewandte mathematik* 40 (1850): 51–88.
7. Thomson, Sir William, and Tait, P. G. *Treatise on Natural Philosophy,* vol. 1. Oxford: Clarendon Press, 1867.
8. Rayleigh, J. W. S. *The Theory of Sound,* vol. 1. New York: Dover Publications, 1945.
9. Kromm, A. "Verallgemeinerte Theorie der Plattenstatik." *Ingenieur-Archive* 21 (1953): 266; "Über die Randquerkräfte bei gestütztn Platten." *Zeitschrift fuer angewandte mathematik und mechanik* 35 (1955): 231.
10. Marcus, S. *Die Theorie elastischer Gewebe.* Berlin: Berlin Press, 1932, p. 12.
11. Navier, C. L. M. H. *Bulletin des Sciences de la Société Philomathique de Paris,* 1823.

12. Levy, Maurice. "Sur l'équilibre élastique d'une plaque rectangular."
    Compt. Rend. 129 (1899).
13. Hildebrand, Francis B. *Advanced Calculus for Applications.* Engle-
    wood Cliffs: Prentice-Hall, Inc., 1963.
14. Peterson, Thurman S. *Elements of Calculus.* New York: Harper and
    Brothers, 1950.
15. Griffel, William. *Plate Formulas.* New York: Frederick Ungar Pub-
    lishing Co., Inc., 1968.
16. Lanczos, Cornelius. *The Variational Principles of Mechanics.* To-
    ronto: University of Toronto Press, 1960.
17. Greenwood, Donald T. *Principles of Dynamics.* Englewood Cliffs:
    Prentice-Hall, Inc., 1965.
18. Dettman, John W. *Mathematical Methods in Physics and Engineer-
    ing.* New York: McGraw-Hill Book Co., 1962.
19. Taylor, Angus E. *Advanced Calculus.* Boston: Ginn and Company,
    1955.
20. Langhaar, Henry L. *Energy Methods in Applied Mechanics.* New
    York: John Wiley and Sons, Inc., 1962.
21. Ritz, W. "On a New Method of Solving Certain Variational Prob-
    lems of Mathematical Physics." *Journal für die Reine and Ange-
    wandte (Crelle)* 135 (1908).
22. Miner, Douglas F., and Seastone, John B., eds. *Handbook of Engi-
    neering Materials,* 1st ed. New York: John Wiley and Sons, Inc.,
    1955.
23. Cauchy, A. L. "De la pression dans un systeme de point materiels."
    Exercises de Mathematique (1828).
24. Lechnitsky, S. G. _____ *(Anisotropic Plates),* 2nd ed.
    Moscow: Govt. Publication of Technical Theoretical Literature,
    1957.
25. Lechnitsky, S. G. *Theory of Elasticity of an Anisotropic Body.* San
    Francisco: Holden-Day, Inc., 1963.
26. Troitsky, M. S. *Orthotropic Bridges Theory and Design.* Cleveland:
    The James F. Lincoln Arc Welding Foundation, 1967.
27. Huffington, H. J. "Theoretical Determinations of Rigidity Proper-
    ties of Orthogonally Stiffened Plates." *Journal of Applied Me-
    chanics* 23 (1956).
28. McFarland, D. E. "An Investigation of the Static Stability of Cor-
    rugated Rectangular Plates Loaded in Pure Shear." Unpublished
    Ph.D. dissertation, University of Kansas, 1967.
29. Salvadori, M. G., and Baron, M. L. *Numerical Methods in Engi-
    neering.* Englewood Cliffs: Prentice-Hall, Inc., 1961.
30. Weeg, G. P., and Reed, G. B. *Introduction to Numerical Analysis.*
    Waltham, Mass.: Blaisdell Publishing Co., 1966.

31. Janssen, T. L. "Static Deflections and Vibrations of Plates by the Method of Ritz Modified by Lagrange Multipliers." Unpublished master's thesis, Wichita State University, 1970.

32. Przemieniecki, J. S. *Theory of Matrix Structural Analysis.* New York: McGraw-Hill Book Co., 1968.

33. Zienkiewicz, O. C., and Cheung, Y. K. *The Finite Element Method in Structural and Continuum Mechanics.* New York: McGraw-Hill Book Co., 1967.

34. Irons, B. M., and Draper, K. J. "Inadequacy of Nodal Connections in a Stiffness Solution for Plate Bending." *AIAA Journal* 3 (1965): 965.

35. Gallagher, R. H. "Analysis of Plate and Shell Structures." *Proceedings of the Symposium on Application of Finite Element Methods in Civil Engineering,* Vanderbilt University, November 13–14, 1969.

36. Clough, R., and Tocher, J. "Finite Element Stiffness Matrices for the Analysis of Plate Bending." *Proceedings of First Conference on Matrix Methods in Structural Mechanics,* Wright Patterson AFB, Ohio, AFFDL TR 66–80, November 1965.

37. Clough, R., and Felippa, C. "A Refined Quadrilateral Element for Analysis of Plate Bending." *Proceedings of Second Conference on Matrix Methods in Structural Mechanics,* Wright Patterson AFB, Ohio, October 1968.

INDEX